About the Author

Shan Gao graduated from the Institute of Electronics, Chinese Academy of Science. He is an independent research scientist in the fields of foundations of quantum theory, quantum superluminal communication and quantum consciousness etc. He is the author of two Chinese books *Quantum Motion and Superluminal Communication* (Monograph) and *Quantum* (Popular book).

E-mail: rg@mail.ie.ac.cn; Homepage: http://www.quantummotion.org

Quantum Motion

Unveiling the Mysterious Quantum World

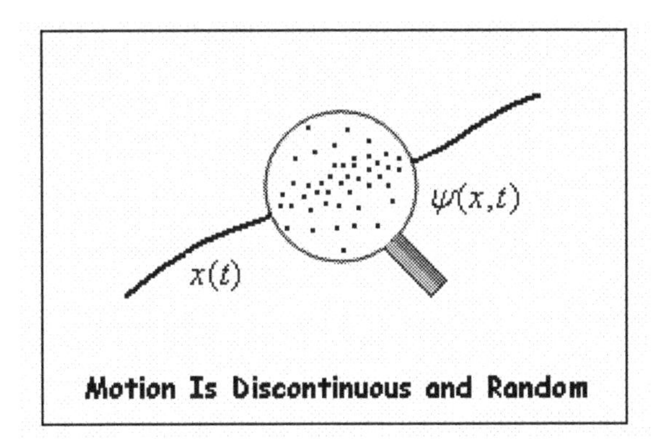

Shan Gao

Published 2006 by arima publishing

www.arimapublishing.com

ISBN 1-84549-148-3
ISBN 978-1-84549-148-2

Printed and bound in the United Kingdom

Typeset in Times New Roman

abramis is an imprint of arima publishing

arima publishing
ASK House, Northgate Avenue
Bury St Edmunds, Suffolk IP32 6BB
t: (+44) 01284 700321

www.arimapublishing.com

To the lonely universe

Contents

Preface

Relativity and quantum theory brought two profound revolutions in 20th century physics. They have become the foundation stones of modern physics. However, these two theories are not only incomplete, but also incompatible. On the one hand, the conceptual foundations of quantum theory have not yet been firmly laid. For example, it is still unknown what on earth the wave function describes and whether or not the wave function really collapses. On the other hand, the existence of quantum non-locality implies that the absolute validity of the principle of relativity will also be challenged. For example, a preferred Lorentz frame may be required for describing the quantum non-local processes. Moreover, general relativity and quantum theory also conflict with each other at the most fundamental level. These facts clearly indicate that our present understandings of space-time and motion require a revolution profounder than that brought by relativity and quantum theory.

This book aims to provide a uniform basis for quantum theory and relativity. Such a basis is indispensable for a unified physical theory. As we know, the basic task of physics is to study the motion of matter in space and time. Then what is the real form of motion? Through a deep analysis of space-time and motion, it is demonstrated that the real motion is the random discontinuous motion of particles in discrete space and time, which is called quantum motion, and the microscopic and macroscopic motions are both displays of quantum motion. As a result, what the wave function describes is quantum motion, and the evolution of quantum motion naturally leads to the dynamical collapse of the wave function. This provides an ontological basis for quantum theory. In addition, it is argued that quantum motion may also explain the maximum and constancy of the speed of light in special relativity, and provide a consistent framework for the unification of quantum theory and general relativity. Consequently, quantum motion may be the uniform basis of quantum theory and relativity. In the last part of this book, the perplexing quantum non-locality is analyzed in detail in terms of quantum motion. It is shown that the collapse of the wave function requires the existence of a preferred Lorentz frame. This provides a natural way to reconcile quantum non-locality and special relativity. A principle of quantum superluminal communication is further introduced through considering the influence of the conscious observer. The analysis also leads to an interesting quantum theory of consciousness.

Reading this book requires a basic knowledge of both relativity and quantum theory. Advanced mastery of these subjects is not necessary. I appeal to the ability to reason rather than the mathematical ability of the reader. An open-minded reader may understand the new ideas in this book more easily. The quantum puzzle may be the most bewildering problem in the history of science; the reader must therefore be prepared to get rid of some ingrained prejudices such as the prejudice of the uniqueness of continuous motion when reading this book. Once these implicit prejudices are rejected, everyone can understand quantum. Although quantum motion may be remote from or even contradict our everyday experience of motion, it is more natural in logic and closer to reality. It is intelligible to everyone. I hope that this book will appeal to all those who have been looking for a real understanding of nature.

The ideas of this book come out of my lonely exploration in the past twenty years. I have also benefited from discussions with many researchers who are interested in the fundamental problems of physics. I want to thank them all, and apologize for not mentioning them by name. At the same time, I am very grateful to my parents Mr. QingFeng Gao and Mrs. LiHua Zhao, my wife HuiXia Liu and my lovely daughter RuiQi Gao. This book could not have been completed without their care and support.

CHAPTER 1

How Do Objects Move?

It is a solid experiential fact that objects can move, and macroscopic objects appear to move continuously. However, direct experience does not tell us how objects move in reality. We cannot simply regard the appearance as the realistic picture. Today it is still a tough task to find how objects move and further understand the motion phenomena.

In this chapter we try to find the realistic picture of motion from the familiar phenomena of motion. It is argued that the phenomena of inertial motion and spontaneous decay imply that motion is spontaneous. The spontaneity of motion requires that motion is essentially random and discontinuous. This conclusion is also supported by the theories of point set and measure in mathematics. However, the randomness and discontinuity of motion cannot emerge in continuous space and time. This is unnatural in logic and inconsistent with experience. We further analyze the motion in discrete space and time. It is shown that the discreteness of space and time not only leads to the existence of random discontinuous motion, but also can naturally release the randomness and discontinuity of motion as experience reveals. Accordingly the real motion may be the random discontinuous motion in discrete space and time.

1.1 Zeno's Paradoxes and the 'at-at' Theory

It is a standard assumption that space and time intervals consist of extensionless points. The moving object is in one position at each instant during the course of motion. But it seems that such a natural assumption does not permit the existence of motion. The famous arrow paradox of Zeno

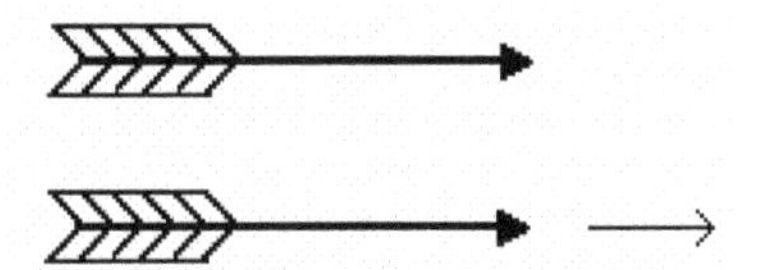

Figure 1 Flying arrow cannot move

provides an interesting argument (cf. Salmon 1970; Papa-Grimaldi 1996). It argues that at any instant a flying arrow cannot move for that would require the instant to have parts, and an instant is by definition a minimal and indivisible element of time. Since at each instant of its flight the arrow is at rest, and time is composed of such instants, the arrow never moves. The standard solution of the paradox rests on what is called the 'at-at' theory of motion. According to the theory, motion is merely a feature of being in different locations at different times, and that is that. As Russell (1903) wrote, "Motion consists merely in the occupation of different places at different times." So it is true that there is no motion during any instant. Motion has nothing at all to do with what happens *during* instants; it has instead to do with what happens *between* instants. If the object has the same location in the instants immediately neighboring, then we say it is at rest; otherwise it is in motion.

However, the 'at-at' theory does not tell us how the different points in space and time intrinsically correlate. In other words, this theory cannot explain dynamism as it never operates the synthesis which could intrinsically correlate different points in space and time. We all know that the transition of different positions is in fact accomplished, but how the transition from one position to another position has been accomplished remains a mystery in the existing theory. In the following, we will try to complete the 'at-at' theory of motion and find how objects move.

1.2 Is Motion Continuous?

How on earth do objects move? Most people may think motion is evidently continuous; this accords with our everyday experience of the motion of macroscopic objects. But is continuous motion the real motion?

We are only accustomed to continuous motion after all, and we cherish it so deeply. We have been taking for granted that continuous motion is the only possible form of motion, as well as the real motion. Indeed, the existence of continuous motion seems to be very natural. An object will hold its velocity if no influences are imposed on it, as there are no causes to change its velocity. Then the free object can only be at rest or move continuously with a constant velocity. In addition, a moving object is in one position at one instant, and it can only be in a neighboring position at the next instant, as there are no causes to result in its sudden appearance in another non-adjacent position. In a similar way, an object cannot move from one position to another position without passing through the in-between positions either, as there is still no cause to result in such a "jump". In a word, the existence of continuous motion is inevitable. If it is not the real motion, then which form of motion is the real motion? If we never see, never learn of and even never dream of another form of motion, how could it be the real motion?

But what is continuous motion? How do we confirm its existence? As we know, an object moving continuously from point 0 to point 1 in a line must pass through all points between 0 and 1. However, there are uncountable points between 0 and 1, say 1/2, 1/4, 1/8 etc. We cannot count up to them during a finite time interval. Then how can we know the object really passes through all these points? If we cannot know that, how can we confirm that the motion of the object is continuous? There may exist some other ways to confirm the existence of continuous motion. For example, although we cannot directly verify that the object passes through all points between 0 and 1, we may confirm it through a plausible hypothesis. One such hypothesis is that an object moving from one position to another position must pass through the middle position. However, even though the existence of continuous motion can be confirmed in terms of such a hypothesis, how can we verify this hypothesis? It may be right for a large distance, but has it been verified for a very small distance? Since there exists uncountable distances, the above hypothesis cannot be verified either. In addition, even if the hypothesis has been

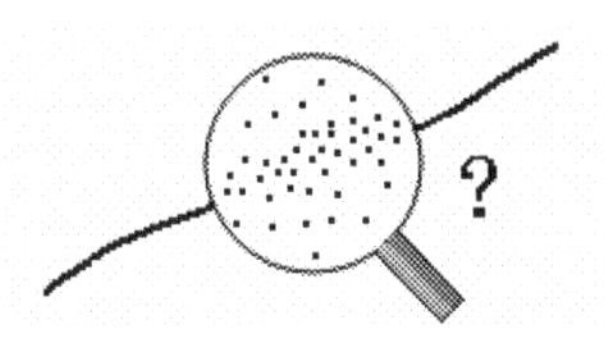

Figure 2 Is motion continuous?

confirmed by experiments, it can only confirm the dense property of the trajectory of the object, and cannot confirm its continuity. For example, the trajectory which is only composed of rational points evidently satisfies the above hypothesis. Accordingly we cannot confirm the existence of continuous motion in terms of such a hypothesis either. In a word, continuous motion is only an assumption or a belief, which can never be confirmed. Moreover, the trajectory of continuous motion, if it exists, can only exist in the meaning of dense point set, since we can never measure a single point or a point set with zero measure in physics.

It appears that infinity prevents us from finding the real motion. In order to find the real motion, we must enter into smaller and smaller space, even the infinitesimal space. Then how far can we walk along the logical road?

1.3 The Spontaneity of Motion

An object will continue to move after it is put into motion. This is the familiar phenomenon of inertial motion[1]. It is well summarized in Newton's first law of motion (i.e. the law of inertial motion) for macroscopic objects. According to the law, a free object can move or change its position, and no external forces are needed to sustain its motion. To our surprise, an in-depth analysis of such ordinary phenomena will lead us to find the real motion.

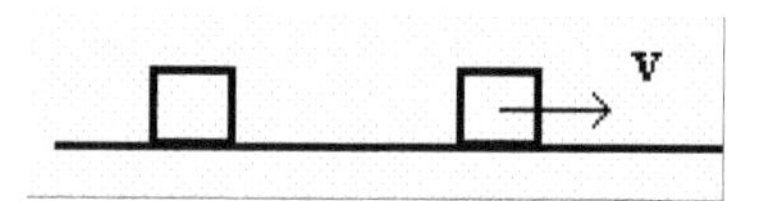

Figure 3 An object can spontaneously change its position

It appears that the inertial motion can be understood in the framework of classical mechanics which assumes objects move in a continuous way. A free object should hold its velocity, since there is no cause leading to the change of its velocity. Thus the object must

[1] Note that pure inertial motion does not occur in nature. According to the existing physical theories, it can only occur at an infinite distance from all sources of gravity.

continuously move in a straight line with a constant speed as the law of inertial motion requires. If such an explanation is complete for the understanding of inertial motion, then the motion of a free object has no spontaneity. It just holds its previous state.

However, a further analysis will show that the motion of a free object may have spontaneity. It is a standard assumption that space and time intervals consist of extensionless points, and a moving object is in one position at each instant during the course of motion. First, the properties defined at instants such as position cannot determine the position change of an object. By virtue of logic and definition, such properties at one instant only contain the information about the object at the instant, and contain no information about the positions of the object at other instants. Second, the properties defined during infinitesimal time intervals such as velocity cannot determine the position change of an object either. In fact, these properties are determined by the position change of the object during an infinitesimal time interval according to their definitions. Consequently, no properties or causes determine the position change of a free object at each instant, and the object must change its position spontaneously during the course of inertial motion (cf. Gao 2001b, 2002a, 2003b).

We stress again that the velocity property, even if it exists, cannot determine the change of the instantaneous position of an object. On the one hand, velocity may not exist for some forms of motion such as Brown motion. Its valid definition requires that motion is continuous and the trajectory is differentiable relative to time. But it is still unclear whether or not the motion of objects is continuous, and thus the continuity of motion and the existence of velocity should not be a precondition when we analyze how objects move. On the other hand, even if velocity may exist for the motion of an object, it does not count as part of the instantaneous state of the object (cf. Albert 2000; Arntzenius 2000; Butterfield 2005). Thus velocity cannot determine the change of the instantaneous state such as position of the object. The orthodox definition of velocity is that the instantaneous velocity for an object is the limit of the object's average velocity as the time-interval around the point in question tends to zero.

As a result, the orthodox velocity is local but temporally extrinsic (cf. Butterfield 2005). The object's instantaneous velocity at an instant codes a lot of information about what its velocity and location are at nearby times, which is given precisely by the limit definition of velocity. In a word, even if the orthodox velocity exists for the motion of an object, it is not an instantaneous intrinsic property of the object in essence, and thus velocity cannot determine the change of the instantaneous position of the object.

There exists two possible ways to avoid the spontaneity of motion. One way is to assume space and time consist of no smallest sized intervals such as points. Rather, space and time are infinitely divisible. The other way is to assume the state of an object at an instant does include a velocity. Such velocity is not defined in terms of the position development of an object during a time interval. Rather, it is a primitive intrinsic feature of an object at an instant, which causes the object to subsequently move in the direction in which the intrinsic velocity is pointing. There are some detailed discussions of these non-standard assumptions recently (cf. Tooley 1988; Vallentyne 1997; Albert 2000; Arntzenius 2000, 2003, 2004; Lewis 2001; Smith 2003; Floyd 2003; Butterfield 2005). If one of these assumptions is right, then the motion of objects will have no spontaneity, and the above explanation provided by classical mechanics may be complete for the understanding of the inertial phenomena. On the other hand, if motion is indeed spontaneous, then these assumptions will be wrong, i.e., space and time consist of smallest sized intervals, and there exists no intrinsic velocity that determines the change of the position of an object in each smallest sized time interval.

There are more direct evidences of the spontaneity of motion in the microscopic world. For example, the alpha particles can spontaneously move out from the radioactive isotopes without any external cause. Such a phenomenon is well known as radioactivity or spontaneous decay, which widely exists in the microscopic world. During the spontaneous decay process, the decay time of each radioactive atom in the substance is completely random. Such randomness also indicates that the decay process happens without causes, and it is spontaneous. In addition, there also exists spontaneous motion during the process of interaction between particles. According to quantum field theory, the interaction between particles is transferred by the transfer particles. Since there are no other transfer particles or interactions between the interacting particles and the transfer particles, the transfer particles must move spontaneously in the process.

The existence of the spontaneous motion without cause seems very counterintuitive. However, it may have a deeper logical basis (cf. Gao 2001b, 2002a, 2003b). If motions can exist only as a result of a certain cause such as interaction between particles, the particles would not be able to move without such interaction, but, on the other hand, the interaction cannot exist if there are no moving particles to transfer it. Thus either everything is immobile or there exist uncaused, spontaneous motions. In short, if the particles can not move in a spontaneous way, then all interactions will not exist, and all particles will also be resting. Furthermore, since the properties of a particle such as mass depend on its interaction with other particles, the particles will be devoid of any properties, and will not exist either. Thus it seems that objects must move spontaneously in order to exist. This means that the existent need of objects may be taken as the universal cause for the spontaneous motion of objects. This kind of cause is independent of each concrete motion process[2].

---

[2] In order to further understand the conclusion, it is necessary to distinguish two kinds of causes. One is the concrete cause which relates to concrete motion process, and the other is the universal cause which is irrelevant to concrete motion process. The former is our familiar cause appearing in the principle of causality. Such a concrete cause will result in a lawful change at a concrete time. The latter is a new kind of cause, which is not included in the existing principle of causality. It may result in ceaseless spontaneous changes (as we will see below, such changes are random). As a consequence, both lawful changes and random changes have their causes. We may call this conclusion the generalized principle of causality. According to the principle, there are two kinds of causes: concrete causes and universal causes, and accordingly there are two kinds of events: lawful events and random events.

1.4 Motion Is Discontinuous and Random

The evidences of spontaneous motion strongly favour the standard assumption that space and time consist of smallest sized intervals such as 0-sized points, and there exists no intrinsic velocity to determine the change of the position of an object in each smallest sized time interval. In such space and time, a moving object is in one position at one instant, and spontaneously moves to another position at another instant. Then how is the transition from one position to another position accomplished? Or how does the object move?

We first consider the motion of a free object. According to the above analysis, a free object can spontaneously change its position. The spontaneity of free motion means that no causes determine the position change of the free object. Here we give a summary of the no-causes argument. (1) There are no internal causes such as an intrinsic velocity to determine the position change of the free object. The object at one instant has no velocity to hold for determining the position of the object at the next instant. The object cannot hold its previous position either. It must move during the course of inertial motion. (2) There are no external causes such as the influences of other objects to determine the position change of the free object either. The free object moves without any external influence. Thus we conclude that no causes determine the position change of the free object.

Since no causes determine the position change of a free object, and the change without cause should be random, the position change of the free object will be random in nature. Whereas the change of position is random at any instant, the trajectory must be discontinuous everywhere. Accordingly the motion of free objects should be essentially discontinuous and random (cf. Gao 1993, 2001b, 2002a, 2003b). It should be stressed once again that the free object has no velocity to hold for determining the change of its instantaneous position. Thus the free object really does not know which direction to move along, and must move in a random and discontinuous way.

We then consider the motion of an object interacting with other objects. Can the interaction determine the position of the object at each instant and change the random discontinuous motion (RDM) to deterministic continuous motion (DCM)? The answer is negative. The essential reason lies in that a completely random process cannot be changed to a deterministic process in principle. If the interaction is not random, then it is evident that the motion of an object under such influence will still be random. If the interaction is also random, then since the combination of two random processes still leads to a random process, the motion of an object under such influence will also be random. Accordingly the motion of an object interacting with other objects is still discontinuous and random. Moreover, the mechanism of interaction may even require the existence of RDM. As we have argued, the interaction between particles is transferred by the transfer particles, and the transfer particles move spontaneously during the course of interaction. Since the spontaneous motion without cause should be discontinuous and random, the motion of the transfer particles will be discontinuous and random.

The existence of RDM may also be justifiable from a mathematical point of view (cf. Gao 1999c, 2000, 2001a, 2002a, 2006a). As we know, the motion state of an object is not the instantaneous state, but the infinitesimal interval state in continuous space and time. An infinitesimal time interval contains uncountable instants. This indicates that the motion state of an object should be described by a point set in space and time, in which the point represents the center of mass of the object at one instant. According to the point set theory in mathematics, the general point set in continuous space and time is a random discontinuous point set. Since we have no *a priori* reason to assume a special point set such as a continuous line for the motion state of an object, the point set describing the motion state should be a random discontinuous point set in space and time (see Figure 4). Thus the object will always move in a discontinuous and random way during an infinitesimal interval at any instant.

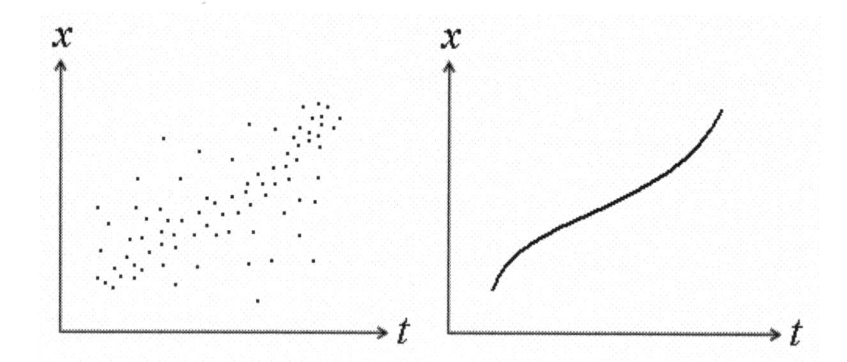

Figure 4 Discontinuous motion and continuous motion: It is really a wonder that so many points bind together to form a continuous curve in order.

To sum up, objects must move in a discontinuous and random way in space and time, which consist of smallest sized intervals such as 0-sized points. In such space and time, continuous motion is impossible in logic. During the motion, the transition from one position to another position is discontinuous and random, and there is no correlation between the different points in space and time at all. This will complete the 'at-at' theory of motion.

1.5 Double-slit Experiment[3]

Concerning the strange discontinuous motion, even those open-minded people may hardly accept it. This is very natural, since it contradicts our everyday experience of the motion of macroscopic objects. However, if you would like to take an objective attitude, you may also think motion is probably not continuous for very small objects, which cannot be directly observed by our naked eyes. Now let's come back to the domain of experience to see whether some phenomena or experiments have revealed discontinuous motion.

---

[3] See Section 5.6 for a detailed discussion.

The double-slit experiment may be one of the experiments which could agree with what you think, since it cannot be explained in terms of the assumption of continuous motion. In the double-slit experiment, a single particle such as an electron is emitted from the source one at a time, and then passes through the two slits to finally arrive at the screen. When a large number of particles reach the screen, they collectively form a double-slit interference pattern.

According to the assumption of continuous motion, the single particle can only pass through one of the two slits. One expects that the double-slit interference pattern should be the same as the direct mixture of two one-slit patterns, each of which is formed by opening each of the two slits independently. The reason is that the passing process of each particle in a double-slit experiment is exactly the same as that in one of the one-slit experiments. However, the results of experiment are that the interference patterns of the above two situations are very different. Thus a single particle must pass through both slits in the double-slit experiment, and its motion will be discontinuous.

Up to now, the double-slit experiment has been accomplished for many kinds of microscopic particles such as electrons. Accordingly we find that the motion of small objects is really discontinuous.

1.6 Quantum Motion

We have been analyzing the motion of objects in continuous space and time, in which space and time consist of 0-sized points. However, the appearance of infinity in quantum field theory and singularity in general relativity has implied that space and time may be not continuous but discrete. In fact, it has been widely argued that the proper combination of quantum theory and general relativity may inevitably result in the discreteness of space and time. In this section, we will analyze the motion of objects in discrete space and time.

In discrete space and time, space and time consist of smallest finite-sized intervals, i.e., there exist a minimum time interval T_U and a minimum space interval L_U. As a result, a particle is no longer in one position at one instant (as in continuous space and time), but limited to a space unit L_U during a time unit T_U in discrete space and time. This defines the existent form of a particle in discrete space and time.

The analysis of motion in continuous space and time also applies to the motion in discrete space and time. In addition, the discreteness of space and time has more restrictions on the possible forms of motion. As we will see, the discreteness of space and time may also require the existence of RDM. Due to the limitation of discreteness of space and time, there are only two possible free motion states for continuous motion: one is rest state, the other is the motion state with a constant speed $c \equiv L_U / T_U$. If the speed of an object is larger than c, the object will move more than a space unit during a time unit. Then moving a space unit will correspond to a time interval shorter than the time unit during such movement. This contradicts the above definition of discrete space and time. On the other hand, if the speed of an object is smaller than c, the object will move a space unit during a time interval longer than the time unit. Then the object will move a space interval shorter than the space unit during a time unit during such movement. This also contradicts the above definition of discrete space and time. Thus a free object can only be still or move with the constant speed c in discrete space and time. This result is evidently inconsistent with experience. A free object can move with a speed different from c in reality[4]. Thus if space and time are indeed discrete as defined

---

[4] It can be conceived that the free object moves with the speed c during some time units, and stays still during the other time units. The average speed of such motion can be different from the speed c, and thus such motion can be consistent with the existing experience. However, the speed change of the free object during such motion can hardly be explained. In addition, such motion will contain some kind of unnatural randomness (e.g. during each time unit the speed of the free object will assume c or zero in a random way),

above, the motion of objects must not be continuous, but be discontinuous and random. This means that an object moving from one space unit to another space unit must not pass the in-between space units. It is just in a space unit during a time unit, and is in another space unit during another time unit.

RDM can naturally exist in discrete space and time. In fact, it may exist only in discrete space and time. As we know, the discontinuity and randomness of motion is absorbed into the motion state defined during an infinitesimal time interval in continuous space and time. As a result, the evolution law of the motion state will be essentially a deterministic continuous equation. Then how can the randomness and discontinuity emerge? And how can the spontaneity of motion present itself? If space and time are continuous, then the inherent randomness of motion cannot be released. Since the 0-sized instants have no physical effects, the randomness and discontinuity of motion cannot emerge through detectable physical effects. This result seems very unnatural in logic, and contradicts one of our most basic scientific beliefs, the minimum ontology. According to the principle, existence should display itself. If a certain thing does exist, then it can be detected, whereas if a certain thing cannot be detected essentially, then it does not exist. Furthermore, if the randomness and discontinuity of motion cannot emerge, the spontaneity of motion cannot present itself either. This is also inconsistent with the evidences of the spontaneity of motion in the microscopic world. Certainly, we can assume there exists other possible sources of randomness, which revise the continuous evolution equation by adding a random evolution term. This can be consistent with the existing experience. However, the existence of RDM still contradicts the minimum ontology. In addition, assuming two different kinds of randomness may not satisfy the requirement of Occam's razor. By comparison, it is more natural and simpler to assume the inherent randomness of motion can emerge and generate the actual randomness and spontaneity of

---

and the randomness has no logical basis either.

motion. Such process can happen in discrete space and time. There exists a minimum time interval T_U in discrete space and time. In contrast to the 0-sized instants, the minimum finite-sized intervals can have a physical effect. Concretely speaking, the object undergoing RDM stochastically stays in a space unit L_U during a time unit T_U, and such random stay can have a small random effect on the evolution of RDM due to the finite duration of the stay. Then during a longer time interval, such small random effect can continually accumulate to generate the detectable randomness and spontaneity of motion.

In a word, we show that space and time may be actually discrete, and the real motion of objects may be the RDM in discrete space and time.

1.7 A Unified Road to Reality

The basic task of physics is to study the motion of matter in space and time. What is the form of space and time? And what is the form of motion? People have been trying to find the answers of these questions. In this book, we will give a possible answer in terms of the existing experience and theories, and further realize the unification of the basis of modern physics.

We postulate that space and time are discrete, motion is discontinuous and random, and the RDM in discrete space and time is the basic form of motion. On the basis of this postulate, a unified theory of physics can be uniquely formulated by three parameters: space unit, time unit and unit of motion. All physical quantities will be expressed by the combinations of them. The theory will contain no constants. This is the beauty of unification. In the following, we will briefly introduce the basic framework of the theory.

1. Postulates

Space and time are discrete, and their quanta are respectively denoted by the space unit L_U and the time unit T_U. Motion is discontinuous and random, and its quantum is denoted by the unit of motion $\hbar$. The RDM in discrete space and time is called quantum motion.

The existing physical theories suggest that L_U and T_U are respectively the double of the Planck length and the Planck time, and $\hbar$ is the Planck constant divided by 2π. When assuming the international unit system, the values of L_U and T_U are approximately $3.2\times10^{-35}m$ and $1.1\times10^{-43}s$, and the value of $\hbar$ is approximately $1.1\times10^{-34}J\cdot s$. In principle, their values can all assume the unit of number 1. Such unit system is called basic unit system.

2. Deductions

2.1 Dynamical wave function collapse and quantum theory

In the non-relativistic domain, the evolution equation of quantum motion includes two parts: the linear Schrödinger evolution term and the non-linear stochastic evolution term. The former corresponds to Schrödinger's equation in quantum mechanics, and the latter describes the dynamical collapse of the wave function. This provides a complete quantum theory.

2.2 The constancy of the speed of light and special relativity

The speed of light can be expressed by the space unit L_U and the time unit T_U, namely $c = L_U / T_U$. The discreteness of space and time will lead to the maximum and constancy of the speed of light. This provides a logical foundation for special relativity. As a result, special relativity will be replaced by the theory of special relativity in discrete space and time.

2.3 The gravitational constant and general relativity

The gravitational constant can be expressed by the space unit L_U, the time unit T_U and the unit of motion $\hbar$, namely $\kappa \equiv 8\pi G/c^4 = 2\pi L_U T_U/\hbar$. This expression implies that gravity may exist only in discrete space and time. As a result, general relativity will be replaced by the theory of general relativity in discrete space and time.

2.4 The unification of quantum and gravity

The conflict between quantum theory and general relativity can be settled with the help of quantum motion. As a result, the theory of quantum motion may provide a consistent framework for the unification of quantum and gravity.

2.5 Charges and interactions

A basic charge can be constructed by using the space unit L_U, the time unit T_U and the unit of motion $\hbar$, namely $C_I = \sqrt{E_U \cdot L_U} = \sqrt{\hbar L_U/T_U} = \sqrt{\hbar c}$, and its interaction potential energy is in inverse proportion to the distance between the charges. The charge of the electromagnetic interaction can be expressed as a multiple of the basic charge, namely $e = \sqrt{\alpha \cdot \hbar c} = \sqrt{\alpha} \cdot C_I$, where $\alpha \approx 1/137$ is the fine structure constant. Similarly, the charges of weak interaction and strong interaction can also be expressed as a certain multiples of the basic charge. The multiple denotes the coupling strength of the interaction, and may be determined by the interplay of the charge and the actual vacuum.

2.6 Dark energy and the fate of our universe

The dark energy may originate from the quantum fluctuations of the discrete space-time limited in our universe. The real form of dark energy will determine the future of our universe.

2.7 Unit of mass and the essential elements of matter

A unit of mass can be expressed by the space unit L_U, the time unit T_U and the unit of motion $\hbar$, namely $M_U = \hbar/(L_U c) = \hbar T_U / L_U^2$. The size of a particle with a unit of mass is the space unit L_U. Considering the size limitation of discrete space-time, the fundamental particles with a unit of mass may be the essential elements of matter.

Most of the above deductions will be detailedly discussed in the following chapters. It can be seen that space-time, motion, interaction and structure of matter will synthesize a uniform picture of nature on the basis of quantum motion. It has a fetching beauty. Let's now enjoy it!

CHAPTER 2

Motion in Continuous Space and Time

In this chapter we will give a detailed analysis of the random discontinuous motion (RDM) in continuous space and time. It is shown that the wave function in quantum mechanics is a mathematical complex describing the motion, and the simplest non-relativistic evolution equation of such motion is just Schrödinger's equation in quantum mechanics. This strongly implies that what (linear) quantum mechanics describes is the RDM in continuous space and time.

2.1 The Physical Definition

A strict definition of the RDM in continuous space and time can be given using three presuppositions about the relation between the physical motion and the mathematical point set. The definition is:

(1). Space and time are both continuous.

(2). A particle is represented by one point in space and time.

(3). The RDM of a particle is represented by a random discontinuous point set in space and time[5].

---

[5] A random discontinuous point set is defined as a set of points (t,x) in continuous space and time, for which the function $x(t)$ is discontinuous and random at all instants. The definition of a discontinuous function is as follows. Suppose A is an open set in $\Re$ (say an interval $A=(a,b)$, or $A=\Re$), and $f:A\rightarrow\Re$ is a function. Then f is discontinuous at $x\in A$, if f is not continuous at x. Note

The first presupposition defines the continuity of space and time. The second one defines the existent form of a particle in continuous space and time. The last one defines the RDM of a particle using the mathematical point set. Here the usual classical world-line of a particle is replaced by a more general world-set. The physical picture of RDM is as follows. A particle is in one position at each instant. The particle moves throughout the whole space with a certain position distribution density during an infinitesimal time interval. The trajectory of the particle is not continuous, but discontinuous and random everywhere.

2.2 The Mathematical Description

The mathematical description of RDM can be obtained by using the measure theory which was first founded by the French mathematician Lebesgue in 1901 (cf. Morgan 1989; Cohn 1993; Nielsen 1994). According to the measure theory, the basic property of a random discontinuous point set, which describes the RDM of a particle, is the measure of the point set. By comparison, the basic property of a continuous line, which describes the continuous motion of a particle, is the length of the line. In the following, we will give the mathematical description of RDM. For simplicity, but without losing generality, we will mainly analyze the one-dimensional motion in space and time which corresponds to the point set in two-dimensional space and time.

---

that a function $f: A \rightarrow \Re$ is continuous if and only if for every $x \in A$ and every real number $\varepsilon > 0$, there exists a real number $\delta > 0$ such that whenever a point $z \in A$ has distance less than δ to x, the point $f(z) \in \Re$ has distance less than ε to $f(x)$.

We first analyze the mathematical description of the RDM of a single particle. Consider the motion state of a single particle in finite intervals Δt and Δx near a space-time point (t_i, x_j) as shown in Figure 5.

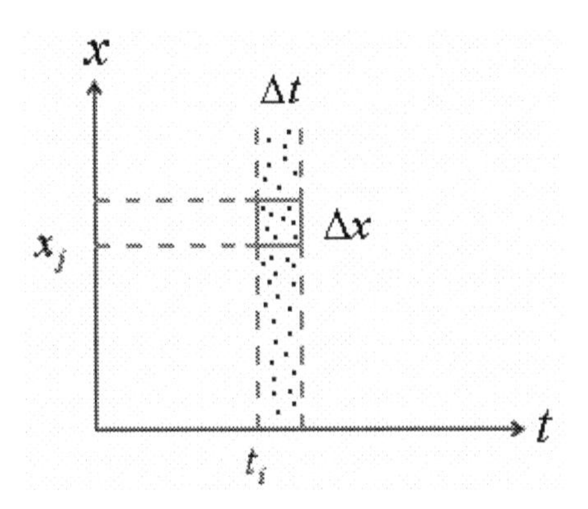

Figure 5 The description of the RDM of a single particle

According to the above definition of RDM, the position of the particle forms a random discontinuous point set in the whole space for the time interval Δt near the instant t_i. Accordingly, there is a local discontinuous point set in the space interval Δx near the position x_j. The local discontinuous point set represents the motion state of the particle in the finite intervals Δt and Δx near the space-time point (t_i, x_j). We study its projection in the *t*-axis, namely the corresponding dense instant set in the time interval Δt. Let *W* be the discontinuous trajectory or world-set of the particle and Q be the square region $[x_j, x_j + \Delta x] \times [t_i, t_i + \Delta t]$. The dense instant set can be denoted by $\pi_t(W \cap Q) \subset \Re$, where π_t is the projection on the *t*-axis. According to the measure theory, we can define the Lebesgue measure:

$$M_{\Delta x, \Delta t}(x_j, t_i) = \int_{\pi_t(W \cap Q) \subset \Re} dt \tag{2.1}$$

Since the sum of the measures of all such dense instant sets in the time interval Δt is equal to the length of the continuous time interval Δt, we have:

$$\sum_j M_{\Delta x,\Delta t}(x_j, t_i) = \Delta t \tag{2.2}$$

Then we can define the measure density:

$$\rho(x,t) = \lim_{\substack{\Delta x \to 0 \\ \Delta t \to 0}} M_{\Delta x,\Delta t}(x,t) / (\Delta x \cdot \Delta t) \tag{2.3}$$

The limit exists for a random discontinuous point set. This provides a strict mathematical description of the point distribution situation for the above local discontinuous point set. We call this measure density position measure density.

Since the local discontinuous point set represents the motion state of the particle, the position measure density $\rho(x,t)$ will be a descriptive quantity of the RDM for a single particle. It represents the relative frequency of the particle appearing in an infinitesimal space interval dx near position x during an infinitesimal interval dt near instant t. From the relation (2.2) we can see that $\rho(x,t)$ satisfies the normalization relation, namely $\int_{-\infty}^{+\infty} \rho(x,t) dx = 1$. Furthermore, we can define the position measure flux density $j(x,t)$ through the relation $j(x,t) = \rho(x,t) v(x,t)$, where $v(x,t)$ is the velocity of the local discontinuous point set. Due to the conservation of measure, $\rho(x,t)$ and $j(x,t)$ satisfy the following equation:

$$\frac{\partial \rho(x,t)}{\partial t} + \frac{\partial j(x,t)}{\partial x} = 0 \tag{2.4}$$

The position measure density $\rho(x,t)$ and the position measure flux density $j(x,t)$ provide a complete description of the RDM of a single particle.

It is very natural to extend the description of the motion of a single particle to the motion of many particles. For the RDM state of N particles, we can define a joint position measure density $\rho(x_1, x_2, ... x_N, t)$. This represents the relative probability of the situation in which particle 1 is in position x_1, particle 2 is in position x_2, ... , and particle N is in position x_N. In a similar way, we can define the joint position measure flux density $j(x_1, x_2, ... x_N, t)$. It satisfies the joint measure conservation equation:

$$\frac{\partial\rho(x_1, x_2, ... x_N, t)}{\partial t} + \sum_{i=1}^{N} \frac{\partial j(x_1, x_2, ... x_N, t)}{\partial x_i} = 0 \tag{2.5}$$

When these N particles are independent, the joint position measure density $\rho(x_1, x_2, ... x_N, t)$ can be reduced to the direct product of the position measure density of each particle, namely $\rho(x_1, x_2, ... x_N, t) = \prod_{i=1}^{N} \rho(x_i, t)$. It should be noted that the joint position measure density $\rho(x_1, x_2, ... x_N, t)$ and the joint position measure flux density $j(x_1, x_2, ... x_N, t)$ are not defined in the three-dimensional real space, but defined in the 3N-dimensional configuration space. This indicates that there exists some kind of similarity between the given description of RDM and the wave function in quantum mechanics.

2.3 Two Kinds of Descriptions

As to the RDM of a particle, the particle is in one position at each instant, but during an infinitesimal time interval the particle will move throughout the whole space with a certain position distribution density. As a result, locality and non-locality coexist in RDM. Since the motion state is defined during an infinitesimal time interval, there should also exist a non-local description of the motion state besides the local position description $\rho(x,t)$ and $j(x,t)$.

The non-local descriptive quantity will be a new quantity which is different from position. In the following, we will give the non-local description of the RDM of a particle.

It is a reasonable assumption that, for a basic non-local state in which the non-local quantity assumes a definite value, the particle will move throughout the whole space with the same position distribution density. The position distribution density of such a non-local basis will be $\rho(x,t)=1$. Using the relation $j=\rho v$ and the measure conservation requirement we find $j(x,t)=v_0$, where v_0 is a constant velocity. Accordingly the definite value of the non-local quantity must relate to the constant velocity. The relation can be written as $p_0 = mv_0$, where p_0 is a definite value of the non-local quantity denoted by p, and m is a proportional coefficient. We call such non-local quantity momentum and the coefficient mass. This is consistent with our present nomenclatures.

It can be seen that, for the particle with a constant momentum, its position will be not limited in an infinitesimal space interval dx during an infinitesimal time interval dt, but spread throughout the whole space with a constant position measure density. Thus momentum is indeed one kind of non-local descriptive quantity of RDM. By comparison, position is one kind of local descriptive quantity of RDM. Since there exists no velocity for a particle undergoing RDM, its momentum no longer relates to a non-existent velocity of the particle, but rather relates to the velocity of the whole discontinuous position set to some extent. As a result, momentum and position are independent in nature. The existence of two kinds of descriptive quantities is essentially due to the discontinuity of motion. For continuous motion there is only one kind of local description, momentum and position are both local descriptive quantities, in which momentum equals to the local velocity of a particle (i.e. the first derivative of position with respect to time) multiplied by its mass.

Since the position measure density $\rho(x,t)$ is not a constant for a general state of RDM, the non-local description of such a state will be not one momentum, but some kind of momentum distribution. This means that the general momentum (motion) state of a particle during an infinitesimal interval is also a random discontinuous point set in momentum space. Similar to the descriptive quantity position, we can also define the momentum measure density $f(p,t)$, which satisfies the normalization relation $\int_{-\infty}^{+\infty} f(p,t)dp = 1$[6], and the momentum measure flux density $J(p,t)$. They satisfy the similar measure conservation equation $\frac{\partial f(p,t)}{\partial t} + \frac{\partial J(p,t)}{\partial p} = 0$. As a simple but typical example, the momentum description is $f(p,t) = \delta^2(p - p_0)$ and $J(p,t) = 0$ for the above momentum basis with one momentum p_0, whose position description is $\rho(x,t) = 1$ and $j(x,t) = p_0 / m$, where $f(p,t)$ satisfies the normalization relation $\int_{-\infty}^{+\infty} f(p,t)dp = \int_{-\infty}^{+\infty} \rho(x,t)dx$. Similarly, for the position basis, whose position description is $\rho(x,t) = \delta^2(x - x_0)$ and $j(x,t) = 0$, the corresponding momentum description is $f(p,t) = 1$ and $J(p,t) = J_0$, where J_0 is determined by x_0. Note that the symmetry between the position description and the momentum description essentially results from the dialectic relation between locality and non-locality coexisting in RDM. They not only are opposite each other, but also embody one another. The non-local basis is composed of the local bases, while the local basis is composed of the non-local bases.

---

[6] For the situation where the integrals $\int_{-\infty}^{+\infty} \rho(x,t)dx$ and $\int_{-\infty}^{+\infty} f(p,t)dp$ are both infinite, the normalization relation will be $\int_{-\infty}^{+\infty} \rho(x,t)dx = \int_{-\infty}^{+\infty} f(p,t)dp$.

Now we have two kinds of descriptive quantities: one is position, the other is momentum. The position description $\rho(x,t)$ and $j(x,t)$ provides a complete local description of the motion state. This we may call the local description of RDM. Similarly the momentum description $f(p,t)$ and $J(p,t)$ provides a complete non-local description of the motion state. This we may call the non-local description of RDM. Since at any instant the motion state of a particle is unique, these two kinds of descriptions should be equivalent, and there should exist a one-to-one relation between them, i.e., there should exist a one-to-one relation between the position description $\rho(x,t)$, $j(x,t)$ and the momentum description $f(p,t)$, $J(p,t)$, and this relation is irrelevant to the specific motion state.

In the following, we will try to find the one-to-one relation between the position description and the momentum description. As we will see, the relation can further determine the possible evolution equations of motion.

2.4 A Heuristic Derivation of the Schrödinger Equation

It can be seen that there exists no direct one-to-one relation between the position measure density $\rho(x,t)$ and the momentum measure density $f(p,t)$. For example, we have $\rho(x,t)=1$ and $f(p,t)=\delta^2(p-p_0)$ for the momentum basis, and there is no a one-to-one relation between them. Then in order to find the one-to-one relation, we need to construct two proper integrative descriptions on the basis of the position description $\rho(x,t)$, $j(x,t)$ and the momentum description $f(p,t)$, $J(p,t)$.

We first disregard the time variable t or let $t=0$. For the momentum basis, we have $\rho(x,0)$ =1, $j(x,0)=p_0/m$ and $f(p,0)=\delta^2(p-p_0)$, $J(p,0)$ =0. We need to synthesize a proper position state function $\psi(x,0)$ using 1 and p_0/m and a proper momentum state function $\varphi(p,0)$ using $\delta^2(p-p_0)$ and 0, and find the one-to-one relation between these two state functions. We assume the general form of the relation is:

$$\psi(x,0)=\int_{-\infty}^{+\infty}\varphi(p,0)T(p,x)dp \tag{2.6}$$

where $T(p,x)$ is the transformation function which is continuous and finite for finite p and x. Since there are many kinds of synthesizing ways and correspondingly many kinds of one-to-one relations, we here assume the existence of a linear one-to-one relation between the proper synthesized state functions such as $\psi(x,0)$ and $\varphi(p,0)$. The following analysis will confirm this assumption. In addition, the proper synthesizing forms should be the same for position description and momentum description due to the symmetry between these two kinds of descriptions.

Since the function $\varphi(p,0)$ should contain a certain form of the element $\delta^2(p-p_0)$, we may generally expand $\varphi(p,0)$ as $\varphi(p,0)=\sum_{i=1}^{\infty}a_i\delta^i(p-p_0)$. In addition, the function $\psi(x,0)$ should contain the constant p_0, and should be continuous and finite for finite x in general. Since the terms besides the term $\delta(p-p_0)$ will lead to infiniteness, the function $\varphi(p,0)$ can only contain the term $\delta(p-p_0)$, and thus we get $\varphi(p,0)=\delta(p-p_0)$. Whereas the function $\varphi(p,0)$ is generally complex, the simplest integrative forms should be $f(p,0)=|\varphi(p,0)|^2$ and $\rho(x,0)=|\psi(x,0)|^2$. Note that the integrative forms $f(p,0)=\varphi^2(p,0)$ and $\rho(x,0)=\psi^2(x,0)$ permit no existence of a one-to-one relation.

The form of $\psi(x,0)$ can be further obtained through analyzing the wave-like property of the momentum basis. Consider the motion state with two different momenta p_0 and $-p_0$, whose momentum measure density is the same 1/2 for p_0 and $-p_0$. We try to find the corresponding position description of this state. First, the position measure density $\rho(x,0)$ cannot be a constant. If the position measure density is a constant, then the state of motion can only contain one momentum due to the limitation of one-to-one relation, i.e., it must be a momentum basis. This result has revealed some kind of superposition property of RDM. For the motion state with two momenta, its position measure density is not the direct mixture of the position measure densities of the motion states with each of these two momenta. Secondly, since the momentum description only contains one parameter, the position description should also contain one parameter due to the existence of one-to-one relation between them. There are two possible forms for the position measure density: one is a nonperiodic space function concentrating in a local region, whose characteristic width W_0 is determined by p_0, the other is a periodic space function extending the whole space, whose period L_0 is determined by p_0. Since when $p_0 \to 0$, $\rho(x,0) \to 1$, the characteristic width W_0 and the period L_0 are both inversely proportional to p_0. In addition, when $p_0 \to \infty$, $W_0 \to 0$, and the first form of position measure density will approach a δ function such as $\delta^2(x)$, whose corresponding momentum measure density is $f(p,t)=1$, thus the first form is excluded. Accordingly the position measure density $\rho(x,0)$ can only be a periodic space function extending the whole space, whose space period L_0 is inversely proportional to p_0. It likes one kind of standing wave. Since each momentum basis has a constant position measure density distributing throughout the whole space, it may be very natural that the superposition of two momentum bases has a periodic position measure density distributing throughout the whole space, whose period is determined by the momenta in superposition.

The above analysis implies that the proper state of motion with one momentum p_0 must also contain a space period L_0 in some way, and should be like one kind of traveling wave, although such periodicity is not revealed in the position measure density and the position measure flux density. Considering the relation $\rho(x,0)=|\psi(x,0)|^2$, the simplest form of the momentum basis will be $\psi(x,0)=Ae^{ip_0x/\hbar}$, where $A=\rho^{1/2}(x,0)$ and $\hbar$ is a constant quantity with dimension $J\cdot s$. Its space period is $L_0=\hbar/p_0$. Just as the superposition of two traveling waves spreading in opposite directions forms a standing wave, the above state with two momenta in opposite directions indeed has a static periodic position measure density. For simplicity we let $\hbar=1$ in the following discussions unless state otherwise.

Now we can get the simplest one-to-one relation for momentum bases:

$$\psi(x,0)=\int_{-\infty}^{+\infty}\varphi(p,0)e^{ipx}dp \tag{2.7}$$

where $\psi(x,0)=e^{ip_0x}$ and $\varphi(p,0)=\delta(p-p_0)$. This relation mainly results from the essential equivalence and symmetry of the local position description and the non-local momentum description of RDM.

In order to find how the time variable t is included in the functions $\psi(x,t)$ and $\varphi(p,t)$, we need to consider the superposition of two momentum bases:

$$\psi(x,t)=\frac{\sqrt{2}}{2}[e^{ipx-ic(t)}+e^{i(p+\Delta p)x-i[c(t)+\Delta c(t)]}] \tag{2.8}$$

Its position measure density is $\rho(x,t) = \frac{1}{2}[1 + \cos(\Delta c(t) - \Delta px)]$. When $\Delta p \to 0$, we have $\rho(x,t) \to 1$, $j \to \rho \frac{p}{m}$ and $\Delta c(t) \to 0$. Then using the measure conservation relation we can obtain $dc(t) = dp \frac{p}{m} t$. As to the nonrelativistic situation where the mass m is a constant quantity, we have $c(t) = \frac{p^2}{2m} t$. We define $E \equiv \frac{p^2}{2m}$ as the energy of a free particle in the nonrelativistic domain. Thus as to any momentum basis we have the time-included formula $\psi(x,t) = e^{ipx - iEt}$, and the complete one-to-one relation is:

$$\psi(x,t) = \int_{-\infty}^{+\infty} \varphi(p,t) e^{ipx - iEt} dp \tag{2.9}$$

Since the one-to-one relation between the position description and the momentum description is irrelevant to the specific motion state, the above one-to-one relation for the momentum bases should hold true for any motion state.

In fact, there may exist more complex forms for the state functions $\psi(x,t)$ and $\varphi(p,t)$. For example, they are not the above simple number functions, but multidimensional vector functions such as $\psi(x,t) = (\psi_1(x,t), \psi_2(x,t), \dots, \psi_N(x,t))$ and $\varphi(p,t) = (\varphi_1(p,t), \varphi_2(p,t), \dots, \varphi_N(p,t))$. However, the above one-to-one relation still holds true for each component function, and these vector functions still satisfy the above modulo square relations, namely $\rho(x,t) = \sum_{i=1}^{N} |\psi_i(x,t)|^2$ and $f(p,t) = \sum_{i=1}^{N} |\varphi_i(p,t)|^2$. These complex forms will describe the particles with more inner properties such as charge and spin etc. For example, as to the particles with spin 1/2 such as electrons, we have N = 4 and $\rho(x,t)=\sum_{i=1}^{4}|\psi_i(x,t)|^2$.

Now we can finally work out the simplest nonrelativistic evolution equation of RDM. First, as to the momentum basis $\psi(x,t)=e^{ipx-iEt}$, we can find its nonrelativistic evolution equation using the above definition $E\equiv\frac{p^2}{2m}$:

$$i\hbar\frac{\partial\psi(x,t)}{\partial t}=-\frac{\hbar^2}{2m}\frac{\partial^2\psi(x,t)}{\partial^2 x} \tag{2.10}$$

Here we have included the constant quantity $\hbar$. Owing to the linearity of the evolution equation, it also holds true for the linear superpositions of momentum bases, which are all possible motion states due to the completeness of the momentum bases. Alternatively we can say that it is the free evolution equation of RDM.

Secondly, we will consider the evolution equation of RDM under an external potential[7]. Since potential is a classical description of interaction, we need to consider the average display of RDM. In the classical nonrelativistic domain, we have $dp/dt=-\partial U/\partial x$, where $U(x,t)$ is an external potential. As to RDM, this formula is an average form, which can be expressed as $d\langle p\rangle/dt=\langle -\partial U/\partial x\rangle$, where $\langle -\partial U/\partial x\rangle=\int_{-\infty}^{+\infty}(-\partial U/\partial x)\cdot\rho(x,t)dx$, $\langle p\rangle=\int_{-\infty}^{+\infty}p\cdot f(p,t)dp$. In order to satisfy this relation, the evolution equation of RDM under an external potential should be:

[7] Different from the derivation of the free evolution equation, the derivation of the evolution equation in the presence of an external potential is not fundamental. In principle, we should derive the equation from the basic form of interaction. See Chapter 8 for a detailed analysis.

$$i\hbar\frac{\partial\psi(x,t)}{\partial t}=-\frac{\hbar^2}{2m}\frac{\partial^2\psi(x,t)}{\partial^2 x}+U(x,t)\psi(x,t) \tag{2.11}$$

For three-dimensional situation the equation will be

$$i\hbar\frac{\partial\psi(\vec{x},t)}{\partial t}=-\frac{\hbar^2}{2m}\nabla^2\psi(\vec{x},t)+U(\vec{x},t)\psi(\vec{x},t) \tag{2.12}$$

This is the simplest nonrelativistic evolution equation of RDM. It can be seen that it takes the same form of Schrödinger's equation in quantum mechanics. In this meaning, we have derived the Schrödinger equation in quantum mechanics.

2.5 The Wave Function is a Complete Description of the Motion

One of the results of the above analysis is that the state function $\psi(x,t)$ provides a complete description of RDM. According to the above evolution equation, the state function $\psi(x,t)$ can be expressed by the position measure density $\rho(x,t)$ and the position measure flux density $j(x,t)$, namely

$$\psi(x,t)=\rho^{1/2}e^{iS(x,t)/\hbar} \tag{2.13}$$

where $S(x,t)=m\int_{-\infty}^{x}\frac{j(x',t)}{\rho(x',t)}dx'+C_0$. This is actually the proper integrative form of the position description. On the other hand, the position measure density $\rho(x,t)$ and the position measure flux density $j(x,t)$ can also be expressed by the state function $\psi(x,t)$, namely

$$\rho(x,t)=|\psi(x,t)|^2 \tag{2.14}$$

$$j(x,t)=\frac{\hbar}{2im}(\psi^*\frac{\partial\psi}{\partial x}-\psi\frac{\partial\psi^*}{\partial x}) \tag{2.15}$$

Accordingly there exists a one-to-one relation between $\rho(x,t)$, $j(x,t)$ and $\psi(x,t)$ when omitting the absolute phase. Since the position measure density $\rho(x,t)$ and the position measure flux density $j(x,t)$ provide a complete description of RDM, the state function $\psi(x,t)$ also provides a complete description of RDM.

2.6 The Meaning of the Theory

The sameness between the simplest nonrelativistic evolution equation of RDM and Schrödinger's equation in quantum mechanics strongly suggests that what (linear) quantum mechanics describes is RDM. However, before reaching a definite conclusion we need to understand the meaning of the theory of RDM in continuous space and time. This means that we must talk about measurement.

One subtle problem is what happens during a measurement process. There exist only two possibilities. One is that the measurement process still satisfies the above evolution equation of RDM or Schrödinger's equation, the linear superposition of the state function can hold all through. This possibility corresponds to the many-worlds interpretation of quantum mechanics (cf. Everett 1957; DeWitt and Graham 1973). The other is that the measurement process does not satisfy the above evolution equation of RDM or Schrödinger's equation, the linear superposition of the state function is destroyed due to some unknown causes. The resulting process is often called the collapse of the wave function. Certainly, the above two possibilities can be tested in experiments. But unfortunately it is very difficult to distinguish them by using present technology. In the following, we will give a brief theoretical analysis of them.

As to the first possibility, the RDM in continuous space and time provides a physical picture for many-worlds interpretation. The particle discontinuously moves throughout all the parallel worlds during a very small time interval or even an infinitesimal time interval. This clearly shows that these parallel complete worlds exist in the same space and time. At the same time, the measure density of the particle in different worlds, which can be strictly defined for RDM, just provides an objective origin of the measure of different worlds. Thus the visualizing physical picture for the many-worlds is one kind of subtle time-division existence, in which each world occupies one tiny part of the continuous time flow. The occupation way is discontinuous and random in essence, i.e., the whole time flow for each world is a dense instant set, and all these dense time sub-flows constitute a whole continuous time flow. In this meaning, the many-worlds are the most crowded in time!

Even though the above many-worlds picture can exist in a consistent way for the particles and measuring apparatus, a hard problem does appear when considering the conscious observer. Why does the observer continuously perceive only one definite world while he is discontinuously moving throughout the many-worlds? This would appear to be inconsistent with our basic scientific belief that our conscious perception is a correct reflection of the objective world. A theory of consciousness may be needed to solve the observer problem. This seems very unnatural.

As to the second possibility, RDM may provide a natural random source for generating the dynamical collapse of the wave function. In addition, it has been argued that gravity may be the cause of the collapse of the wave function (cf. Penrose 1996), and that the Planck energy or the Planck time may relate to the process (cf. Percival 1994; Hughston 1996; Fivel 1997b; Gao 2000, 2006b). Accordingly the RDM in discrete space and time may provide a physical basis for the dynamical collapse theory.

CHAPTER 3

Motion in Discrete Space and Time

Whereas space and time may be actually discrete, we further analyze the random discontinuous motion (RDM) in discrete space and time. It is shown that the evolution of such motion may naturally lead to the dynamical collapse of the wave function, and this collapse will finally bring the appearance of continuous motion in the macroscopic world. This provides a uniform realistic picture of the microscopic and macroscopic worlds.

3.1 The Existence of Discrete Space and Time

Quantum theory and general relativity are both based on the continuous space-time assumption. However, the appearance of infinity in quantum field theory and singularity in general relativity has implied that space-time may be not continuous but discrete. In fact, it has been widely argued that the proper combination of quantum theory and general relativity may inevitably result in the discreteness of space-time. For example, the formula of black hole entropy has implied that space and time may be discrete. The entropy of a black hole is:

$$S = \frac{A}{4L_P^2} \tag{3.1}$$

where A is the area of the black hole horizon, L_P is the Planck length. In addition, the discreteness of space-time appears more clearly in the generalized uncertainty principle (GUP) (cf. Garay 1995; Adler and Santiago 1999):

$$\Delta x = \Delta x_{QM} + \Delta x_{GR} \geq \frac{\hbar}{2\Delta p} + \frac{2L_p^{\ 2}\Delta p}{\hbar} \tag{3.2}$$

The first term denotes the position uncertainty from quantum mechanics, and the second term denotes the position uncertainty from general relativity. Their combination results in the existence of a minimum finite position uncertainty $\Delta x_{\min} = 2L_p$. In a similar way, the resulting minimum finite time uncertainty is $\Delta t_{\min} = 2T_p$.

The basic physical definition of discrete space-time can be given as follows. In the discrete space and time, there exist a minimum time interval $T_U \equiv 2T_P$ and a minimum space interval $L_U \equiv 2L_P$, where $T_P = (\frac{G\hbar}{c^5})^{1/2}$ and $L_P = (\frac{G\hbar}{c^3})^{1/2}$ are respectively the Planck time and the Planck length. Any physical being can only exist in the space region not smaller than the minimum space unit L_U, and any physical becoming can only happen during a time interval not shorter than the minimum time unit T_U. As a result, there exists no any deeper space-time structure beneath the minimum space-time sizes. Any space and time differences smaller than the minimum length and minimum time interval are in principle undetectable, i.e., the space-times with a difference smaller than the minimum sizes are physically identical[8].

The discreteness of space-time is a very strong restriction. As we will see, it may result in the happening of the dynamical collapse of the wave function.

---

[8] It should be noted that the discreteness of space-time is essentially one kind of quantum property due to the universal existence of quantum fluctuations, and thus the space-times with a difference smaller than the minimum sizes are not absolutely identical, but nearly identical in physics.

3.2 Discreteness of Space and Time May Result in Quantum Collapse

Quantum measurement problem is the fundamental problem of quantum theory. The theory does not tell us how and when the measurement result appears. As Bell (1993) said, the projection postulate is just a makeshift. In this sense, the existing quantum theory is an incomplete description of reality. Therefore it is natural to consider the continuous Schrödinger evolution and the discontinuous wave function collapse as two ideal approximations of a unified evolution process. The new theory describing such unified evolution is generally called revised quantum dynamics or dynamical collapse theory. It has been widely studied in recent years (cf. Ghirardi, Rimini and Weber 1986; Pearle 1989; Diosi 1989; Ghirardi, Pearle and Rimini 1990; Percival 1994; Penrose 1996; Hughston 1996; Gao 2000, 2006b).

An important problem of revised quantum dynamics is the origin of wave function collapse. It may be very natural to guess that the collapse of the wave function is induced by gravity. The reasons include: (1) gravity is the only universal force being present in all physical interactions; (2) gravitational effects grow with the size of the objects concerned, and it is in the context of macroscopic objects that linear superpositions may be violated. The gravity-induced collapse conjecture can be traced back to Feynman (1995). In his Lectures on Gravitation, he considers the philosophical problems in quantizing macroscopic objects and contemplates on a possible breakdown of quantum theory. He said, "I would like to suggest that it is possible that quantum mechanics fails at large distances and for large objects ...it is not inconsistent with what we do know. If this failure of quantum mechanics is connected with gravity, we might speculatively expect this to happen for masses such that $GM^2/\hbar c = 1$, of M near 10^{-5} grams".

Penrose (1996) further strengthened the gravity-induced collapse argument. He argued that the superposition of different space-times is physically improper, and the evolution of such superposition cannot be defined in a consistent way. This requires that a quantum superposition of two space-time geometries, which corresponds to two macroscopically different energy distributions, should collapse after a very short time. Penrose's argument reveals a profound and fundamental conflict between the general covariance principle of general relativity and the superposition principle of quantum mechanics. According to general relativity, there exists one kind of dynamical interrelation between motion and space-time, i.e., the motion of particles is defined in space-time, at the same time, space-time is determined by the motion of particles. Then when we consider the superposition state of different positions of a particle, say position A and position B, one kind of logical inconsistency appears. On the one hand, according to quantum theory, the valid definition of such superposition requires the existence of a definite background space-time, in which position A and position B can be distinguished. On the other hand, according to general relativity, the space-time, including the distinguishability between position A and position B, cannot be predetermined, and must be dynamically determined by the position superposition state. Since the different position states in the superposition determine different space-times, the space-time determined by the whole superposition state is indefinite. Then an essential conflict between quantum theory and general relativity does appear. Penrose believed that this conflict requires that the quantum superposition of different space-times cannot exist in a precise way, and should collapse after a very short time. Thus gravity may indeed be the physical origin of wavefunction collapse.

In this section, we will give a new argument supporting a gravitational role in quantum collapse. It will be demonstrated that the discreteness of space-time, which results from the proper combination of quantum theory and general relativity, may inevitably result in the dynamical collapse of the wave function. Moreover, the minimum sizes of discrete space-time also yield a plausible collapse criterion consistent with experiments. This analysis will reinforce and complete Penrose's argument.

We will first point out one possible deficiency in Penrose's argument (Gao 2006b). As we think, his argument may inevitably fail in continuous space-time. If space-time is continuous, then the space-times with arbitrarily small difference are physically different. Since the quantum superposition of different space-times is physically ill-defined due to the conflict between general relativity and quantum mechanics, such state cannot exist in reality. Thus the quantum superposition of two space-times with a very small difference such as the superposition states of microscopic particles cannot exist either. This does not accord with experience. However, if space-time is discrete, and there exist a minimum time interval and a minimum space interval, then Penrose's argument can be reinforced and thus can succeed. In short, the discreteness of space-time may cure the above deficiency in Penrose's argument. The key point is that two space-times with a difference smaller than the minimum sizes are the same in physics in discrete space-time. Thus their quantum superposition can exist and collapse after a finite time interval. Only the quantum superposition of two space-times with a difference larger than the minimum sizes cannot exist, and should collapse instantaneously. Such dynamical collapse of the wave function can accord with experience.

In order to make our prescription be precise, we need to define the difference between two space-times. As indicated by the generalized uncertainty principle, namely the formula (3.2), the energy difference ΔE corresponds to the space-time difference $\frac{2L_p^{\ 2}\Delta E}{\hbar c}$. Then as to the states in quantum superposition with energy difference ΔE, the difference between the space-times determined by the states may be characterized by the quantity $\frac{2L_p^{\ 2}\Delta E}{\hbar c}$. The physical meaning of such space-time difference can be further clarified as follows. Let the two energy eigenstates in the superposition be limited in the regions with the same radius R (they may locate in different positions). Then the space-time outside the region can be described by the Schwarzschild metric:

$$ds^2 = (1-\frac{r_S}{r})^{-1}dr^2 + r^2 d\theta^2 + r^2 \sin\theta^2 d\phi^2 - (1-\frac{r_S}{r})c^2 dt^2 \tag{3.3}$$

where $r_S = \frac{2GE}{c^4}$ is the Schwarzschild radius. By reasonably assuming that the metric tensor inside the region R is the same as that on the boundary, the proper size of the region is

$$L = 2\int_0^R (1-\frac{r_S}{R})^{-1/2} dr \tag{3.4}$$

Then the space difference of the two space-times in the superposition inside the region R can be characterized by

$$\Delta L \approx \int_0^R \frac{\Delta r_S}{R} dr = \Delta r_S = \frac{2L_p^{\ 2}\Delta E}{\hbar c} \tag{3.5}$$

This result is consistent with the generalized uncertainty principle. Accordingly as to the states in quantum superposition, we can define the difference of their corresponding space-times as the difference of the proper spatial sizes of the regions occupied by the states. Such difference represents the fuzziness of the point-by-point identification of the spatial section of the two space-times. As a result, the space-translation operators are not the same for the two space-times. In comparison with Penrose (1996)'s definition of acceleration uncertainty, our definition may be regarded as some kind of position uncertainty in the superposition of space-times. Such uncertainty does not depend on the spatial distance between the states in the superposition. A detailed comparison of them will be given in the next section.

The space-time difference defined above can be rewritten as the following form:

$$\frac{\Delta L}{L_U} \approx \frac{\Delta E}{E_P} \tag{3.6}$$

where $E_P = h / T_P$ is the Planck energy. This relation seems to indicate some kind of equivalence between the energy difference and the difference of space-times defined above. However, it should be stressed that they are not equivalent for general situations. In physics, it is the difference of space-times, not the energy difference in the superposition that results in the dynamical collapse of the wave function. In addition, the proper size of the region occupied by a state is not solely determined by the energy of the state, but determined by the energy distribution of all entangled states. The latter determines the metric tensor inside the region. For example, as to the entangled state $\psi_1\varphi_1 + \psi_2\varphi_2$, the difference of the proper sizes of the regions occupied by the states ψ_1 and ψ_2 is also influenced by the energy distribution of the entangled states φ_1 and φ_2. Some concrete examples will be given in Section 3.4.

Now we can give a collapse criterion in terms of the above analysis. If the difference of the space-times in quantum superposition is equal to the minimum space unit L_U, the superposition state will collapse to one of the definite space-times within the minimum time unit T_U. If the difference of the space-times in quantum superposition is smaller than L_U, the superposition state will collapse after a finite time interval larger than T_U. As a result, the superposition of space-times can only possess a space-time uncertainty smaller than the minimum space unit in discrete space-time. If such uncertainty limit is exceeded, the superposition will collapse to one of the definite space-times instantaneously.

Lastly, we note that the above collapse criterion is also consistent with the requirement of discrete space-time. This can be seen from the analysis of a typical example. Consider a quantum superposition of two energy eigenstates. The initial state is

$$\psi(x,0)=\frac{1}{\sqrt{2}}[\varphi_1(x)+\varphi_2(x)] \tag{3.7}$$

where $\varphi_1(x)$ and $\varphi_2(x)$ are two energy eigenstates with energy eigenvalues E_1 and E_2. According to the linear Schrödinger evolution, we have:

$$\psi(x,t)=\frac{1}{\sqrt{2}}[e^{-iE_1t/\hbar}\varphi_1(x)+e^{-iE_2t/\hbar}\varphi_2(x)] \tag{3.8}$$

and

$$\rho(x,t)=|\psi(x,t)|^2=\frac{1}{2}[\varphi_1^{\,2}(x)+\varphi_2^{\,2}(x)+2\varphi_1(x)\varphi_2(x)\cos(\Delta E/\hbar\cdot t)] \tag{3.9}$$

This result indicates that the position measure density $\rho(x,t)$ will oscillate with a period $T=h/\Delta E$ in each position of space, where $\Delta E=E_2-E_1$ is the energy difference of the energy eigenstates in the superposition. If the energy difference ΔE exceeds the Planck energy E_p, the difference of the corresponding space-times will be larger than the minimum space unit L_U, and the superposition state cannot exist according to the above collapse criterion. This means that the position measure density $\rho(x,t)$ cannot oscillate with a period shorter than the minimum time unit T_U. This result is consistent with the requirement of discrete space-time. In discrete space-time, the minimum time unit T_U is the minimum distinguishable size of time, and no change can happen during a time interval shorter than T_U.

3.3 A Model of Wavefunction Collapse in Discrete Space and Time

It is well known that a chooser and a choice are needed to bring the required dynamical collapse of the wave function (cf. Pearle 1999). According to the above analysis, the choice should be the energy distribution which determines the space-time geometry. Then who is the chooser? In this section, we will try to solve the chooser problem, and propose a model of wavefunction collapse in discrete space-time in terms of the new chooser and choice.

In the usual wavefunction collapse models, the chooser is generally an unknown random classical field. However, such models may have some inherent problems concerning the chooser. For example, when the classical field is quantized, its collapse will need another random classical field. The process is an infinite chain, which is very similar to the von Neumann (1955)'s infinite chains of measurement. As a result, these models cannot explain where the intrinsic randomness originates from. In order to cut the infinite chain, it is more reasonable that the randomness of the collapse process originates from the wave function itself. It has been argued that the probability relating to the wave function is not only the display of the measurement results, but also the objective character of the motion of particles (cf. Bunge 1973; Shimony 1993). Thus it is not unreasonable to assume that the motion of particles described by the wave function is an intrinsic random process. As we have demonstrated, what the wave function describes is indeed the RDM. Then the randomness of the wavefunction collapse process may result from such random motion of particles, i.e., the chooser may be the RDM described by the wave function. In such a model, the motion of particles naturally provides a random source to collapse the wave function describing the motion, and the dynamical collapse of the wave function is just an inherent display of the motion. This point of view is very natural and simple. In the following, we will propose a model of wavefunction collapse in terms of the new chooser.

We first give a general picture of the collapse process resulting from RDM. The particle undergoing RDM stochastically stays in a space unit L_U near a position x during a time unit T_U near an instant t. The probability of the stay satisfies the position measure density $\rho(x,t)$. Such stay near a position x will change the position measure density $\rho(x,t)$ in the position x. Then during a finite time interval much larger than T_U, the position measure density $\rho(x,t)$ will undergo one kind of stochastic collapse evolution resulting from the random stays. If there is an ensemble of particles with the same initial position measure density $\rho(x,0)$, then the collapse evolution result of each particle will be that the particle is in a random position, i.e., the position measure density is one in that position and zero in other positions, and the position probability distribution of the particles in the ensemble satisfies the initial position measure density $\rho(x,0)$.

We then analyze the influence of the RDM on the wave function. As a typical example, we study a simple two-level system whose initial state is

$$|\psi,0>=\sqrt{\mathrm{P}_1(0)}\,|E_1>+\sqrt{\mathrm{P}_2(0)}\,|E_2> \qquad (3.10)$$

where $|E_1>$ and $|E_2>$ are two energy eigenstates with eigenvalues E_1 and E_2, $\mathrm{P}_1(0)$ and $\mathrm{P}_2(0)$ are the corresponding measure densities or probabilities which satisfy the conservation relation $\mathrm{P}_1(0)+\mathrm{P}_2(0)=1$. Since the linear Schrödinger evolution does not change the probability distribution, we can only consider the influence of dynamical collapse on the probability distribution. As to the RDM described by the above state, the energy of the particle assumes E_1 or E_2 in a random way, and the corresponding probability are respectively $\mathrm{P}_1(0)$ and $\mathrm{P}_2(0)$ at the initial instant. In other words, the particle is in state

$|E_1>$ with probability $P_1(0)$, and is in state $|E_2>$ with probability $P_2(0)$ at the initial instant. In discrete space-time, this means that at the initial instant the particle stays in state $|E_1>$ for a time unit T_U with probability $P_1(0)$, and stays in state $|E_2>$ for a time unit T_U with probability $P_2(0)$.

Assume after the particle stays in state $|E_1>$ for a time unit T_U, $P_1(t)$ turns to be

$$P_{11}(t+T_U)=P_1(t)+\Delta P_1 \tag{3.11}$$

where ΔP_1 is a functional of $P_1(t)$. Due to the conservation of probability, $P_2(t)$ correspondingly turns to be

$$P_{21}(t+T_U)=P_2(t)-\Delta P_1 \tag{3.12}$$

The probability of this stay is $p(E_1,t)=P_1(t)$. Similarly, we assume after the particle stays in state $|E_2>$ for a time unit T_U, $P_2(t)$ turns to be

$$P_{22}(t+T_U)=P_2(t)+\Delta P_2 \tag{3.13}$$

$P_1(t)$ correspondingly turns to be

$$P_{12}(t+T_U)=P_1(t)-\Delta P_2 \tag{3.14}$$

The probability of this stay is $p(E_2,t)=P_2(t)$.

We can work out the diagonal density matrix elements of the evolution:

$$\begin{aligned}\rho_{11}(t+T_U)&=\sum_{i=1}^{2}p(E_i,t)\cdot P_{1i}(t+T_U)\\&=P_1(t)[P_1(t)+\Delta P_1]+P_2(t)[P_1(t)-\Delta P_2]\\&=P_1(t)+[P_1(t)\Delta P_1-P_2(t)\Delta P_2]\\&=\rho_{11}(t)+[P_1(t)\Delta P_1-P_2(t)\Delta P_2]\end{aligned} \tag{3.15}$$

$$\rho_{22}(t+T_U)=\rho_{22}(t)+[P_2(t)\Delta P_2-P_1(t)\Delta P_1] \tag{3.16}$$

Since the probability distribution of the collapse results should satisfy the Born's rule in quantum mechanics, we require $\rho_{11}(t+T_U)=\rho_{11}(t)$ and $\rho_{22}(t+T_U)=\rho_{22}(t)$. In such a way, the probability distribution of the collapse results can reveal the actual measure density of the state. This is a natural requirement when considering the validity of measurement. Then we can obtain the following relation:

$$P_1\Delta P_1-P_2\Delta P_2=0 \tag{3.17}$$

Using the conservation relation $P_1+P_2=1$, this relation can be rewritten as follows:

$$\frac{\Delta P_1}{1-P_1}=\frac{\Delta P_2}{1-P_2} \tag{3.18}$$

When the superposition state contains n branches, the above requirement will lead to the following equations set:

$$\begin{cases} \Delta P_1-\sum_{j\neq 1}\frac{P_j\Delta P_j}{1-P_j}=0 \\ \Delta P_2-\sum_{j\neq 2}\frac{P_j\Delta P_j}{1-P_j}=0 \\ \cdot \\ \cdot \\ \cdot \\ \Delta P_n-\sum_{j\neq n}\frac{P_j\Delta P_j}{1-P_j}=0 \end{cases} \tag{3.19}$$

where $\sum_i P_i=1$, and i, j denotes the branch states. Here we assume that the increase ΔP_i of one branch comes from the scale-down of the other branches, where the scale is the probability P_j of each of these branches. By solving this equations set, we find the following relation:

$$\frac{\Delta P_1}{1-P_1} = \frac{\Delta P_2}{1-P_2} = \dots = \frac{\Delta P_n}{1-P_n} = k \tag{3.20}$$

where $k \in [0,1]$ is an undetermined dimensionless quantity. This is an important relation describing the dynamical collapse of the wave function in discrete space-time.

By using the above relation, we can further work out the non-diagonal density matrix elements of the evolution:

$$\begin{aligned}\rho_{12}(t+T_U) &= \sum_{i=1}^{2} p(E_i,t)\cdot\sqrt{P_{1i}(t+T_U)}\sqrt{P_{2i}(t+T_U)} \\ &= P_1(t)\sqrt{P_1(t)+\Delta P_1}\sqrt{P_2(t)-\Delta P_1} + P_2(t)\sqrt{P_1(t)-\Delta P_2}\sqrt{P_2(t)+\Delta P_2} \\ &\approx (1-\frac{1}{4}k^2)\rho_{12}(t)\end{aligned} \tag{3.21}$$

$$\rho_{21}(t+T_U) \approx (1-\frac{1}{4}k^2)\rho_{21}(t) \tag{3.22}$$

When assuming k is approximately a constant quantity during the collapse process we have:

$$\rho_{12}(t) \approx [1-\frac{1}{4}k^2]^{t/T_U}\ \rho_{12}(0) \tag{3.23}$$

Let $\rho_{12}(t)=\frac{1}{2}\ \rho_{12}(0)$, we can obtain the appropriate collapse time formula:

$$\tau_c \approx \frac{2}{k^2}T_U \tag{3.24}$$

According to the collapse criterion obtained in the last section, the factor k is a functional of the space-time difference ΔL. When $\Delta L = 0$, collapse never happens, thus we have

$\Delta P_i = 0$, $k = 0$. When $\Delta L = L_U$, collapse happens within the minimum time unit T_U, thus we have $\Delta P_i = 1 - P_i$, $k = 1$. Then when assuming the differentiability of the function $k(\Delta L)$ and considering the dimensional relation we can obtain:

$$k(\Delta L) = \sum_{i=1}^{\infty} k^{(i)}(0) \cdot (\frac{\Delta L}{L_U})^i \tag{3.25}$$

When $k^{(1)}(0) = 0$ and $k^{(2)}(0) \neq 0$, the collapse time $\tau_c \approx (\frac{L_P}{\Delta L})^4 \cdot T_P$ is too long to be consistent with experiments. Thus we can get the factor k in the first rank[9]:

$$k = \frac{\Delta L}{L_U} \tag{3.26}$$

Here we omit the dimensionless constant $k^{(1)}(0)$ which is generally in the order of one. Then the collapse time formula is:

$$\tau_c \approx 2(\frac{L_U}{\Delta L})^2 \cdot T_U \tag{3.27}$$

During the dynamical collapse process, when the particle stays in the state $| E_i >$ for a time unit T_U, $P_i(t)$ turns to be

$$\Delta P_i(t) = \frac{\Delta L}{L_U}[1 - P_i(t)] \tag{3.28}$$

By using the equivalent relation (3.6) for the superposition state of two energy eigenstates, we can rewrite the above formulae as follows:

$$\tau_c \approx \frac{2\hbar E_P}{(\Delta E)^2} \tag{3.29}$$

---

[9] How to determine the final form of k is a left problem.

$$\Delta \mathrm{P_i(t)} = \frac{\Delta E}{E_{\mathrm{P}}}[1 - \mathrm{P_i(t)}] \tag{3.30}$$

For a general energy superposition state, ΔE may be defined as the squired energy uncertainty of the state:

$$\Delta E = [\sum_i P_i (E_i - \overline{E})^2]^{1/2} \tag{3.31}$$

where $\overline{E} = \sum_i P_i E_i$ is the average energy of the state. As a result, the collapse time will generally relate to the initial energy probability distribution of the state.

We give some comments on the above collapse model. First, even though the collapse time formula is the same as that in the energy-driven collapse models (e.g. Hughston (1996)) for some special situations such as the above energy superposition state, our collapse model is essentially different from the energy-driven collapse model. In our model, the choice is the energy distribution, while the choice is the whole energy in the energy-driven collapse models. This has been stressed in the definition of the space-time difference ΔL in the last section. Such difference can also be clearly seen in the common position measuring situation, which will be discussed in the next section. The energy-driven collapse model cannot account for the appearance of definite macroscopic measurement results (cf. Pearle 2004), while our collapse model can do.

Next, we give an analysis of the relation between the above collapse time formula and that proposed by Penrose (1996). In Penrose's gravity-induced collapse model, the collapse time formula is $\tau_{\mathrm{c}} \approx \frac{\hbar}{\Delta E_G}$, where ΔE_G is the gravitational self-energy of the difference between the mass distributions belonging to the two states in the superposition

$$\Delta E_G = \frac{1}{G}\int (\nabla\Phi_2 - \nabla\Phi_1)^2 dx^3 \tag{3.32}$$

where Φ_1 and Φ_2 are the Newtonian gravitational potentials of the two states, and G is Newton's gravitational constant. When the states in the superposition are in the same spatial region with radius R, we have

$$\Delta E_G \approx \frac{G(\Delta E)^2}{c^4 R} \tag{3.33}$$

$$\tau_c \approx \frac{\hbar}{\Delta E_G} \approx \frac{c^4 \hbar R}{G(\Delta E)^2} = \frac{1}{\frac{L_P}{R}(\frac{\Delta E}{E_P})^2} T_P \tag{3.34}$$

In our collapse model this requires $k \approx \sqrt{\frac{L_P}{R}} \frac{\Delta E}{E_P}$. This term comes from some kind of acceleration uncertainty in the superposition of space-times according to Penrose's analysis. By comparison, our choice $k \approx \frac{\Delta E}{E_P}$ comes from the position uncertainty in the superposition of space-times. We may consider the former as a 1/2 rank $O(L_P{}^{1/2})$ correction of the latter. Its existence also implies that a term of zero rank $O(1) \approx \frac{\Delta E}{E_P}$ may exist. In fact, the only existence of Penrose's term may contradict the discreteness of space-time as implied from the analysis in the last section. Since Penrose's term is extremely smaller than ours in most situations where $R >> L_P$, it can be omitted in our collapse model. In addition, we note that Penrose's collapse time formula seems to be not right for the situations where the energy eigenstates in the superposition are in different spatial regions. His formula predicts that such superposition should collapse in the Newtonian limit, but Christian (2001) had reasonably argued that the superposition does not collapse.

Lastly, we stress that the above collapse model has some new reasonable characters, and may have some advantages over the existing collapse models. The assumed collapse evolution in the existing collapse models can be generally written as follows (cf. Pearle 1999):

$$|A,t>=\sqrt{P_1(0)}e^{-\frac{1}{4\lambda t}[B(t)-2\lambda ta_1]^2}|a_1>+\sqrt{P_2(0)}e^{-\frac{1}{4\lambda t}[B(t)-2\lambda ta_2]^2}|a_2> \qquad (3.35)$$

where $B(t)$ is a classical Brownian motion function, and λ is a parameter determining the collapse rate. The probability density of $B(t)$ is:

$$P_t(B)\equiv<A,t|A,t>=P_1(0)\ e^{-\frac{1}{2\lambda t}[B(t)-2\lambda ta_1]^2}+P_2(0)\ e^{-\frac{1}{2\lambda t}[B(t)-2\lambda ta_2]^2} \qquad (3.36)$$

First, in our collapse model the chooser is not an unknown random classical field, but the RDM described by the wave function itself. The normal linear evolution and the dynamical collapse of the wave function form a complete evolution of the RDM. This seems more natural and simpler. Secondly, the dynamical collapse process proceeds gradually all the time in our collapse model. Especially when the collapse process approaches completion, the RDM still changes the whole probability distribution gradually. Whereas in the existing collapse models, the change of the whole probability distribution resulting from the assumed noise turns to be very large when the collapse process approaches completion, although such change happens with a very small probability. This way of change seems very unnatural. In this meaning, the collapse dynamics in our model can be regarded as an improved version of Pearle's gambler's ruin game dynamics (cf. Pearle 1999), which is the basis of the existing collapse models. Thirdly, the dynamical collapse equation can be uniquely determined by the RDM in our collapse model. The uniqueness of the collapse law may imply the validity of the model. Lastly, our model is essentially discrete, and has no corresponding formulation in continuous space-time. Its validity strongly relies on the discreteness of space-time, which is an inevitable result of the proper combination of quantum theory and general relativity.

3.4 Some Considerations of the Consistency with Experiments

In our collapse model, the preferred bases are the energy eigenstates, namely the stationary solutions of Schrödinger's equation. Such states correspond to the definite space-time geometries. This is not inconsistent with the microscopic experiments. Even though there is a large spatial spreading for each energy eigenstate, their superposition may have very small spatial spreading. Since the energy uncertainty of such superposition can be very small, its collapse time will be very long. Thus the quantum state with small spatial spreading can still hold throughout the duration of usual experiments. For example, as to a quantum state with spatial spreading $\Delta x \approx 0.1\mu m$, its energy uncertainty can be as small as $\Delta E \approx 1eV$ when satisfying Heisenberg's uncertainty relation, and the collapse time is $\tau_c \approx 10^{12} s$.

In addition, our collapse model does not contradict the macroscopic experiences either. The environmental influence will result in a large energy uncertainty in the quantum superposition of localized states, and thus collapse the superposition to one of the localized states very soon. As a result, the macroscopic objects can always be localized due to the environmental influence (cf. Adler 2001). For example, for a common object of size $10^{-8} cm$ in the atmosphere at standard temperature and pressure, one nitrogen molecule accretes in the object during a time interval of $10^{-8} s$ in average (cf. Redhead 1996; Adler 2001). Thus the energy uncertainty resulting from the accretion fluctuation is $\Delta E \approx 28GeV$ (corresponding to the mass of a nitrogen molecule) for a superposition of two localized states of the object separated by the distance $10^{-8} cm$, and such superposition will collapse to one of the localized states after a time $\tau_c \approx 10^{-8} s$.

In the following, we analyze a typical position measurement experiment in terms of our collapse model. Consider an initial state which describes a particle in the superposition of two locations (e.g. a superposition of two gaussian wavepackets separated by a certain distance). After the usual measurement interaction, the position measuring apparatus evolves to a superposition of two macroscopically distinguishable states:

$$(c_1 \psi_1 + c_2 \psi_2)\varphi_0 \rightarrow c_1 \psi_1\varphi_1 + c_2 \psi_2\varphi_2 \tag{3.37}$$

where ψ_1, ψ_2 are the states of the particle in different locations, φ_0 is the initial state of the position measuring apparatus, and φ_1, φ_2 are the different outcome states of the apparatus. For an ideal measurement, the two particle/apparatus states $\psi_1\varphi_1$ and $\psi_2\varphi_2$ have precisely the same energy spectrum (cf. Pearle 2004). However, since different measurement results appear in different positions of the apparatus, the two particle/apparatus states do possess different energy distribution. For example, different position states of a photon in a superposition are detected in different positions of a photographic plate, and they interact with different AgCl molecules in these positions. Thus we should rewrite the apparatus states as $\varphi_0 = \chi_A(0)\chi_B(0)$, $\varphi_1 = \chi_A(1)\chi_B(0)$, $\varphi_2 = \chi_A(0)\chi_B(1)$, where $\chi_A(0)$ and $\chi_B(0)$ respectively denote the initial states of the apparatus in positions *A* and *B*, $\chi_A(1)$ and $\chi_B(1)$ respectively denote the outcome states of the apparatus in positions *A* and *B*. Such description clearly shows that different outcome states of the apparatus possess different energy distributions. Then we have

$$(c_1\psi_1 + c_2\psi_2)\chi_A(0)\chi_B(0) \rightarrow c_1\psi_1\chi_A(1)\chi_B(0) + c_2\psi_2\chi_A(0)\chi_B(1) \tag{3.38}$$

Since there always exists a certain measurement amplification from the microscopic state to the macroscopic outcome in common measurement process, there will be a large energy difference between the states $\chi_A(0)$, $\chi_B(0)$ and $\chi_A(1)$, $\chi_B(1)$. This means that the apparatus states in the superposition possess very different energy distributions in positions *A* and *B*, and the space-times in the superposition are also very different in these positions. Such difference will result in the proper quantum collapse in the measurement process according to our collapse model. As a typical example, for a photon detector such as avalanche photodiode, its energy consumption is sharply peaked in the very short measuring interval (cf. Berg 1996). One type of avalanche photodiode operates at 10^5 cps and has a mean power dissipation of $4mW$ (cf. Cova et al 1996; Berg 1996). This corresponds to an energy consumption of about $2.5\times10^{11}eV$ per measuring interval $10^{-5}s$. By using the collapse time formula $\tau_c \approx \frac{\hbar E_P}{(\Delta E)^2}$, where the energy difference ΔE between the states in the superposition such as $\chi_A(0)$ and $\chi_A(1)$ is $\Delta E \approx 2.5\times10^{11}eV$, we can work out the collapse time $\tau_c \approx 1.25\times10^{-10}s$. This time size is smaller than and close to the measuring interval. Thus our collapse model is consistent with experiments, and can account for the appearance of definite macroscopic measurement results. In addition, the measurement parameters of avalanche photodiodes may have provided an indirect confirmation of the model.

3.5 From Quantum Motion to Classical Motion: the Unification of Two Worlds

If the motion of objects is essentially discontinuous and random, then why does the motion of macroscopic objects appear continuous? In order to provide a uniform picture of the microscopic and macroscopic worlds, we must answer how the transition from quantum motion to classical motion happens.

The physical picture of quantum motion is as follows. A particle stays in a space unit L_U during a time unit T_U. Then it will still stay there or stochastically appear in another space unit L_U, which may be very far from the original region, during the next time unit T_U.

During a time interval much larger than the time unit T_U, the particle will move throughout the whole space with a certain average position measure density $\rho(x,t)$.

Although the quantum motion of a particle is completely discontinuous and random, the discontinuity of motion is absorbed into the motion state of the particle, which is defined during an infinitesimal time interval in continuous space-time, by the descriptive quantities of position measure density $\rho(x,t)$ and position measure flux density $j(x,t)$. As a result, the evolution law for the motion state of a particle is also a deterministic continuous equation such as Schrödinger's equation in continuous space-time. In discrete space-time, the position measure density $\rho(x,t)$ will undergo one kind of discrete stochastic evolution besides the continuous deterministic evolution, which results in the dynamical collapse of the wave function. However, such stochastic evolution may proceed very soon for a macroscopic object, and its position measure density $\rho(x,t)$ may always concentrate in a very small local region, then the macroscopic object will always be in a local region, and can only be approximately still or in continuous motion. In the following, we will present a more detailed analysis in terms of the evolution law of quantum motion.

Although the complete evolution law of quantum motion is not available now, we can give its general characters according to our previous analysis. The nonrelativistic evolution equation of quantum motion will be a revised Schrödinger equation which contains two kinds of evolution terms. The first is the deterministic linear Schrödinger evolution term, and the second is the stochastic nonlinear evolution term resulting in the dynamical collapse of the wave function. The equation can be formally written in a discrete form:

$$\psi(x,t+T_U)-\psi(x,t)=\frac{1}{i\hbar}H\psi(x,t)T_U+S\psi(x,t) \tag{3.39}$$

where the first term in the right side is the linear Schrödinger evolution term, H is the corresponding Hamiltonian, and the second term in the right side is the stochastic nonlinear evolution term, S is the corresponding stochastic evolution operator. We stress that the equation should be essentially a discrete one in physics, and all quantities are defined in discrete space-time.

In the evolution equation of quantum motion, the linear Schrödinger term will lead to the spreading process of the wave function as in quantum mechanics, while the nonlinear stochastic term will lead to the collapse process or localizing process of the wave function. Accordingly the evolution of quantum motion will be a certain combination of the spreading process and the localizing process. According to the analysis in Section 3.2, the relative strength of the spreading process and the localizing process is mainly determined by the space-time difference or energy distribution difference between different branches of the wave function. If the energy distribution difference is very small, then the evolution will be mainly dominated by the spreading process. This is just what happens in the microscopic world. A particle can pass through the two slits in the double-slit experiment.

If the energy distribution difference between different branches of the wave function is very large as for macroscopic objects[10], then the linear spreading of the wave function will be greatly suppressed, and the evolution of the wave function will be dominated by the localizing process. Such localizing process proceeds continually, and the position measure density $\rho(x,t)$ will always concentrate in a very small local region. Thus a macroscopic object will always be in a local region, and can only be approximately still or in continuous motion. This

---

[10] The largeness of the energy distribution difference for macroscopic objects results mainly from environmental influences such as thermal energy fluctuations. For example, for a macroscopic object comprising 10^{26} atoms, the energy distribution difference resulting from thermal energy fluctuations is $\Delta E \approx N^{1/2}kT \approx 300GeV$ for $T = 300K$, and the collapse time is $\tau_c \approx 10^{-10}s$.

is just the appearance of continuous motion in the macroscopic world. In addition, the space interval which we can perceive is much larger than the spreading region of the position measure density of a macroscopic object. Especially, our perception time is also much longer than the typical stochastic evolution time of a macroscopic object. Thus what we perceive in the macroscopic world also appears to be a continuous flux.

Furthermore, the evolution equation of continuous motion may also be derived from that of quantum motion as an approximation (cf. Gao 2000, 2001a, 2002a, 2006a). Here we only give a simple explanation using the Ehrenfest theorem in quantum mechanics (cf. Schiff 1968), which can be formulated as follows:

$$d\langle x\rangle / dt = \langle p\rangle / m \tag{3.40}$$

$$d\langle p\rangle / dt = \langle -\partial U / \partial x\rangle \tag{3.41}$$

As we have argued, the position measure density will no longer spread for a macroscopic object, thus the average terms in the above formulae will represent the effective descriptive quantities for the continuous motion of a macroscopic object. Then the evolution equation of continuous motion can be naturally derived in such a way. The result is: $dx / dt = p / m$, the definition of momentum, and $dp / dt = -\partial U / \partial x$, the motion equation. We note that there should also exist some other stochastic terms in the above equations, which stem from the stochastic nonlinear evolution term in the equation of quantum motion. Although these terms may be very small for most situations, they may be detected by the more precise experiments.

In a word, quantum motion provides a uniform realistic picture for the microscopic and macroscopic worlds. The most familiar continuous motion is only its approximate display in the macroscopic world.

CHAPTER 4

The Confirmation of Quantum Motion

The existent can be found, and its law can be confirmed. This is our basic scientific belief. If quantum motion is the real motion of matter, its law should permit it to be found and its law to be confirmed. In this chapter we will study the confirmability of quantum motion. The analysis of the measurement of quantum motion will show that its existence and evolution law can indeed be found and confirmed.

4.1 Quantum Entanglement

Since measurement is one kind of physical interaction between the observed system and the measuring apparatus, we must first analyze and understand the entanglement between quantum motions, which is the basis of quantum measurement. In this section, we will analyze the subtle characters of quantum entanglement. For simplicity, but without losing generality, we will mainly discuss the two-particle entangled state.

As we know, the descriptive quantities of the quantum motion of two particles are the joint position measure density $\rho(x_1, x_2, t)$ and the joint position measure flux density $j(x_1, x_2, t)$. Similar to the derivation of the Schrödinger equation of one particle, we can also find the one-to-one relation between the position description and the momentum description and derive the motion equation for the quantum motion of two particles. The one-to-one relation is the following two-fold Fourier transformation:

$$\psi(x_1, x_2, t) = \int_{-\infty}^{+\infty}\int_{-\infty}^{+\infty} \varphi(p_1, p_2, t)\, e^{i(p_1 x_1 + p_2 x_2)}\, dp_1 dp_2 \tag{4.1}$$

The motion equation of two particles is:

$$i\hbar\frac{\partial\psi(x_1,x_2,t)}{\partial t}=-[\frac{\hbar^2}{2m_1}\frac{\partial^2\psi(x_1,x_2,t)}{\partial^2 x_1}+\frac{\hbar^2}{2m_2}\frac{\partial^2\psi(x_1,x_2,t)}{\partial^2 x_2}]+U(x_1,x_2,t)\psi(x_1,x_2,t)$$

(4.2)

It is evident that when two particles are independent, the wave function $\psi(x_1,x_2,t)$ can be reduced to the product of the wave functions of two particles, namely $\psi(x_1,x_2,t)=\psi_1(x,t)\ \varphi_1(x,t)$. Here the motion equation of two particles will be simply reduced to two motion equations of one particle. Then the solution $\psi_1(x,t)\ \varphi_1(x,t)$ is a trivial solution of the above motion equation of two particles. This is the only situation where we can understand the quantum motion of two particles by using one particle picture. Owing to the linearity of the motion equation of two particles, the linear superposition of the above trivial solutions such as $\psi_1(x,t)\ \varphi_1(x,t)+\psi_2(x,t)\ \varphi_2(x,t)$ is still a solution of the motion equation of two particles. Such state of two particles is called two-particle entangled state (TPES).

It can be shown that TPES can be formed through the linear evolution of the product states of two independent particles. Assume the initial state of particle 1 is $\psi(x)$, and the initial state of particle 2 is $\varphi_1(x)+\varphi_2(x)$. The interaction Hamiltonian is $H_I=g(t)A\cdot P$, where $g(t)$ is a smooth function in the interval [0,*T*], $g(0)=g(T)=0$, A is the operator acting on particle 2, and satisfies the relations: $A\ \varphi_1=a_1\ \varphi_1$, $A\ \varphi_2=a_2\ \varphi_2$, P is the momentum operator acting on particle 1. When the interaction interval *T* is so small, the motion of the two particles is mainly determined by the interaction Hamiltonian H_I. Then after the interaction the state of the two particles turns to be:

$$\psi(x-a_1)\,\varphi_1(x)+\psi(x-a_2)\,\varphi_2(x) \tag{4.3}$$

This is a TPES. It is one kind of position-entangled state of two particles. The above forming process of TPES can also be physically explained in terms of the picture of quantum motion. Before particles 1 and 2 begin to interact with each other, particle 1 is in state $\psi(x)$, and particle 2 discontinuously moves throughout the states $\varphi_1(x)$ and $\varphi_2(x)$ with the same measure density during a very small time interval. When particles 1 and 2 begin to interact with each other, due to the discontinuity or time-division property of quantum motion, particle 2 will be in state $\varphi_1(x)$ during some time units, and change the initial state $\psi(x)$ of particle 1 into the time-division state $\psi(x-a_1)$ through the interaction with particle 1, while during the other time units particle 2 will be in state $\varphi_2(x)$, and change the initial state $\psi(x)$ of particle 1 into the time-division state $\psi(x-a_2)$ through the interaction with particle 1. Then the time-division property of quantum motion is transferred from particle 2 to particle 1 through the interaction between them, and this forms the TPES $\psi(x-a_1)\,\varphi_1(x)+\psi(x-a_2)\,\varphi_2(x)$.

Certainly, the above method is just a simple and intuitionistic method to form a TPES. There exist many other delicate methods to form the TPES, one of which is the well-known spontaneous parameter down conversion (SPDC). However, even though the existence of TPES has been confirmed in both theory and experiment, we have not grasped its physical nature yet. In the following, we will further analyze and understand the TPES in terms of the physical picture of quantum motion.

We first give an intuitionistic physical picture of TPES. For a TPES $\psi_1\varphi_1+\psi_2\varphi_2$, particles 1 and 2 are in state $\psi_1\varphi_1$ during a time unit T_U, then they will still stay in this state or be in state $\psi_2\varphi_2$ in a random way during the next time unit T_U. Particles 1 and 2 being in state $\psi_1\varphi_1$ means that particle 1 is in state ψ_1 and particle 2 is in state φ_1. Similarly, particles 1 and 2 being in state $\psi_2\varphi_2$ means that particle 1 is in state ψ_2 and particle 2 is in state φ_2. During a very short time interval which is still much longer than the time unit T_U, the two particles will move throughout the states $\psi_1\varphi_1$ and $\psi_2\varphi_2$ with the same measure density.

Now TPES has revealed its perplexing character, i.e., that there exists one kind of mysterious synchronism between the particles in TPES. Concretely speaking, during any time unit T_U, when particle 1 is in state ψ_1, particle 2 must be in state φ_1, and when particle 1 changes its state to ψ_2, particle 2 will change its state to φ_2 synchronously. These two particles in the TPES will be synchronously in the states ψ_1 and φ_1 or ψ_2 and φ_2 in such way, and this kind of synchronism is irrelevant to the distance between them. Moreover, particles 1 and 2 are in state $\psi_1\varphi_1$ or $\psi_2\varphi_2$ in a random way during each time unit T_U. This is an essential character of quantum motion. Thus it is more difficult to understand this kind of stochastic synchronism.

In addition, there exists one kind of more mysterious "perception at a distance" between the particles in the TPES. When one of the particles in the TPES is measured, the other particle, which may be far away from the measured particle, will instantaneously "perceive" the influence on the first particle. In general, the TPES $\psi_1\varphi_1 + \psi_2\varphi_2$ may have different expansion forms in different state bases spaces. Assume there are two equivalent expansion forms $\psi_1\varphi_1 + \psi_2\varphi_2 = \psi_1'\varphi_1' + \psi_2'\varphi_2'$, where ψ_i, φ_i and ψ_i', φ_i' are different state bases. Then when the measuring apparatus is set to measure the first kind of state bases $\{\psi_i\}$, particle 1 will collapse to ψ_1' or ψ_2' after measurement. Here particle 2 will instantaneously "perceive" the change of particle 1, and synchronously collapses to φ_1 or φ_2. Similarly, when the measuring apparatus is set to measure the second kind of state bases $\{\psi_i'\}$, particle 1 will collapse to ψ_1' or ψ_2' after measurement. Here particle 2 will also instantaneously "perceive" the change of particle 1, and synchronously collapses to φ_1' or φ_2'.

It can be reasonably guessed that these perplexing or even incomprehensible characters of TPES may relate to the weird nature of quantum motion, especially its wholeness. If we try to understand the displays of TPES only using the motion picture of parts, we will always ask how the particles in TPES can hold the stochastic synchronism, and how particle 2 instantaneously "perceives" the change of particle 1 etc. However, we can never find the answers, and there exist no answers at all. In fact, the above questions about TPES are improper. Once the independent particles form a TPES, the concept of part is no longer valid in physics. As to the TPES, there exist no motion states of individual particles. The particles in the TPES form an indivisible whole, and there exists only a motion state of the whole. We call it quantum whole. Thus the above questions about how the parts in the whole hold the synchronism and instantaneously "perceive" one another are improper and meaningless. In the following, we will further analyze the concept of quantum wholeness, and re-understand TPES using it.

First, the existence of quantum whole is primary, while the existence of separable parts is only derivative. We cannot explain the existence of TPES in terms of the existence of independent parts. There exist no separable parts in the TPES, and there is only an indivisible quantum whole there. In addition, as to the situation where separable parts may exist, we can still regard it as a trivial situation of the existence of quantum whole.

Secondly, once a quantum whole forms, no interaction is required to hold its existence. No matter how far the parts are separated, this kind of quantum wholeness will not be impaired. This property of quantum whole is essentially different from that of classical whole. Interaction is required to hold the existence of classical whole. The synchronization between the parts in a classical whole is not instantaneous, and must reckon in the transmission delay of the interaction determined by the distance between them. In addition, the classical wholeness will be weakened if the interaction between the classical parts turns to be weak. When the distance between the classical parts is large enough, the existence of separable parts will be approximately valid in physics. In fact, as to a classical whole, the separable parts can always exist during the time interval when they are in the space-like separating regions.

Thirdly, when we realize that what a TPES describes is a quantum whole, the TPES will possess the same substantiality and understandability as the state of a single particle. They both describe the quantum motion of a certain existence. For the TPES, the object in motion is the quantum whole consisting of two inseparable particles. It can be regarded as one compound particle to some extent, which is composed of two particles separated in space. For the state of a single particle, the object in motion is a single particle. However, it can still be regarded as some kind of quantum whole when the particle has an inner structure.

Fourthly, the existence of quantum whole is not absolute for TPES. It is actually relative to the entangled property of particles. The particles in a TPES may be different kinds, thus they can still be identified. It is still valid to talk about the respective existence of the particles in a TPES in physics.

Lastly, we want to stress that the existence of TPES is a precondition of the measurability of quantum motion. In order to measure a quantum state such as $\varphi_1 + \varphi_2$, we must entangle it with the state of the measuring apparatus ψ. Since the measuring apparatus should be able to generate different result states for different measured states in the superposition, the measurement will naturally require the existence of the TPES such as $\psi_1\varphi_1 + \psi_2\varphi_2$. Thus the above analysis of TPES is undoubtedly indispensable to the study of the measurement of quantum motion.

4.2 The Measurement of Quantum Motion

In this section, the measurement of quantum motion will be detailedly analyzed. We will first discuss the general requirements of a valid measurement of quantum motion, and then demonstrate that the law of quantum motion ensures that these requirements can be satisfied. In this way, we can prove that the existence and the evolution law of quantum motion can be found and confirmed by experiments.

Consider the position measurement of a particle in a stationary state $\psi(x)$. On the one hand, a valid measurement should be able to reveal the real state of the observed system as accurately as possible. As to the particle in the state $\psi(x)$, its real state is that the particle moves throughout the whole space with the position measure density $|\psi(x)|^2$. Thus the position measurement of the particle should be able to reveal this kind of position distribution state, and measure its position measure density $|\psi(x)|^2$. On the other hand, the macroscopic measuring apparatus can only generate a definite measurement result. Then a single position measurement can only measure one definite position of the particle, and cannot completely reveal the position distribution state $\psi(x)$ of the particle. Accordingly a large number of similar measurements are needed to reveal the quantum state of the particle, and the distribution of the measurement results should be able to reflect the position measure density $|\psi(x)|^2$.

There are two alternatives, one is taking a large number of similar measurements of the same particle, and the other is measuring a large number of particles in the same state. We will first demonstrate that the first alternative is impossible for an unknown state. Assume the initial pointer state of the measuring apparatus is $\varphi(x)$, which is a very narrow Gaussian wavepacket centralizing on the position 0. In order to measure the particle in a definite position state $\delta(x-x_1)$, the measuring apparatus must interact with the particle, and the interaction must lead to the corresponding displacement of its pointer, i.e., the pointer state must turn to be $\varphi(x-x_1)$ after the measurement. In this way, we can find the position of the particle in any finite position state $\delta(x-x_i)$ by reading the pointer in the state $\varphi(x-x_i)$. Now we use this measuring apparatus to measure the superposition state of different positions of the particle such as $\delta(x-x_1)+\delta(x-x_2)$, where the distance between x_1 and x_2 is much larger than the average spreading size of the pointer state. Then during the measurement their states are entangled, and the process can be described as follows

$$[\delta(x-x_1)+\delta(x-x_2)]\,\varphi(x) \rightarrow \delta(x-x_1)\varphi(x-x_1)+\delta(x-x_2)\varphi(x-x_2) \qquad (4.4)$$

However, if the measuring apparatus is a valid one, the measurement result can only be a definite value, i.e., the pointer can only be localized in a very narrow space interval. In addition, a valid measurement must reflect the observed state as accurately as possible. Then the result of the above measurement can only be x_1 or x_2, i.e., the pointer state must turn to be $\varphi(x-x_1)$ or $\varphi(x-x_2)$ after such measurement. Accordingly the state of the measured particle will turn to be $\delta(x-x_1)$ or $\delta(x-x_2)$ after the measurement. Thus we find that the measurement of an unknown state of particle will destroy the measured state, and change it to the eigenstate corresponding to the measured result, and thus the following measurements can no longer reveal the original state. This proves that a large number of similar measurements of the same particle cannot reveal an unknown quantum state of particle.

Now there is only the second alternative, namely measuring a large number of particles in the same state $\psi(x)$. In order to find the unknown state $\psi(x)$ of each particle, in which the particle moves throughout the whole space with the position measure density $|\psi(x)|^2$, the distribution of the position measurement results must reflect the position measure density $|\psi(x)|^2$. Moreover, when the number of the measured particles is infinite, they must be exactly the same. Then we get the general characters and requirements of the measurement of quantum motion: (1). A single measurement can only generate one definite result, which partially reflects the measured state. The measurement will generally destroy the measured state, and change it to the eigenstate corresponding to the measured result. (2). A large number of measurements of the particles in the same state are needed, and the distribution of the measurement results should equal to the real measure density of the measured property in the state when the number of particles is infinite.

In the following, we will demonstrate that the law of quantum motion indeed satisfies the above requirements. Let the state of the measured particle be $\alpha^{1/2}\varphi_1(x)+\beta^{1/2}\varphi_2(x)$. The measured property is denoted by operator A, which satisfies the following relations $A\,\varphi_1=a_1\,\varphi_1$ and $A\,\varphi_2=a_2\,\varphi_2$. The initial pointer state of the measuring apparatus is $\psi(x)$. It is a very narrow Gaussian wavepacket centralizing on position 0, whose average width w satisfies $w<<|a_1-a_2|$. The interaction Hamiltonian is $H_I=g(t)A\cdot P$, where $g(t)$ is a smooth function in the interval [0,*T*], $g(0)=g(T)$=0, and P is a momentum operator acting on the measured particle. Then when the interaction interval T is very small, the evolution of the whole system is mainly determined by the interaction Hamiltonian H_I, and the free Hamiltonian and the nonlinear stochastic item in the evolution equation of quantum motion can both be omitted. By solving the motion equation we can get the entangled state:

$$\alpha^{1/2}\,\psi(x-a_1)\varphi_1(x)+\beta^{1/2}\,\psi(x-a_2)\varphi_2(x) \qquad (4.5)$$

where $\psi(x-a_1)$ and $\psi(x-a_2)$ are the Gaussian wavepackets centralizing on positions a_1 and a_2, which average widths are still w. This is the first stage of measurement. We call it state entanglement stage.

After the interaction, the whole system begins to evolve freely according to the evolution equation of quantum motion. This is the second stage of measurement. Since the Schrödinger linear term in the evolution equation does not influence the generation of the measurement result, we may only consider the nonlinear stochastic evolution item. The entanglement of the measuring apparatus with the measured particle will introduce a very large energy distribution difference ΔE between the branches $\psi(x-a_1)\varphi_1(x)$ and $\psi(x-a_2)\varphi_2(x)$. Then according to the collapse law of quantum motion (see Section 3.3), the whole system will stochastically collapse to the branch $\psi(x-a_1)\varphi_1(x)$ or $\psi(x-a_2)\varphi_2(x)$ after a very short interval $\tau_c \approx \frac{2\hbar E_P}{(\Delta E)^2}$, and the corresponding collapse probability are respectively α and β. Here the pointer state has collapsed to the state $\psi(x-a_1)$ or $\psi(x-a_2)$, which indicates that the measurement result is a_1 or a_2. At the same time, the measured state has been destroyed, and collapses to the state $\varphi_1(x)$ or $\varphi_2(x)$. We call the second stage of measurement state collapse stage[11]. Then when a large number of particles in the same state $\alpha^{1/2}\varphi_1(x,t)+\beta^{1/2}\varphi_2(x,t)$ are measured, the measurement results distribution will be approximately $P(a)=\alpha\delta_{aa_1}+\beta\delta_{aa_2}$, which is the real measure density of the measured property A in the state. When the number of particles is infinite, the results distribution will equal to the real measure density.

In a word, we have demonstrated that the law of quantum motion can ensure that the measurement of quantum motion possesses the required characters. Thus the law of quantum motion indeed permits it to be found and its law to be confirmed. This not only warrants the rationality of the existence of quantum motion, but also warrants the consistency of the theory of quantum motion.

4.3 Protective Measurement

As we know, the measurement of an unknown state will inevitably result in the collapse of the wave function, and we cannot directly find the real picture of the quantum motion of a single particle through such measurement. Then whether or not is there one kind of measurement which can reveal the quantum motion of a single particle in a partly known state? The answer is positive. It is just the protective measurement proposed by Aharonov et al (Aharonov and Vaidman 1993; Aharonov, Anandan and Vaidman 1993). In the following, we will briefly introduce its basic principle and discuss its application to quantum motion.

Protective measurement aims at measuring the motion state of a single particle by repeated measurements which do not destroy its state. The general method is to let the measured particle be in a non-degenerate eigenstate of the whole Hamiltonian using a suitable

[11] It should be noted that these two stages of measurements cannot be strictly distinguished in reality, and they always proceed simultaneously.

interaction, and then make the measurement adiabatically so that the wave function of the particle neither changes appreciably nor becomes entangled with the measuring apparatus. The suitable interaction is called the protection. It should be stressed that the state of a particle measured by protective measurement is not completely unknown, and we must know before the measurement whether the measured particle is in a nondegenerate energy eigenstate or which energy superposition state it is in. This is needed to determine how to introduce the protective interaction. Certainly, it is also unnecessary to fully know the state before the protective measurement. For example, we only need to know that the particle in a bound potential is in the minimum energy eigenstate. The state can be naturally achieved through the spontaneous transition of particle, and its form is unknown. Although the protective measurement cannot measure the unknown state of a particle, it may reveal the objective motion of a particle in a known state. In the following, we will demonstrate how protective measurement can directly reveal the quantum motion of a particle.

For simplicity, but without losing generality, we consider a particle in a discrete nondegenerate energy eigenstate $\psi(x)$. The protection is natural for this situation, and no additional protective interaction is needed. The interaction Hamiltonian for measuring the value of an observable A_n in the state is:

$$H_I = g(t)P \cdot A_n \tag{4.6}$$

where P denotes the momentum of the pointer of the measuring apparatus, A_n is a normalized projection operator on small regions V_n having volume v_n, which can be written as follows:

$$A_n = \begin{cases} \dfrac{1}{v_n}, & x \in V_n \\ 0, & x \notin V_n \end{cases} \tag{4.7}$$

and the time-dependent coupling $g(t)$ is normalized to $\int_0^T g(t)dt = 1$. We let $g(t) = 1/T$ for most of the time T and assume that $g(t)$ goes to zero gradually before and after the period T to obtain an adiabatic process when $T \to \infty$. The initial state of the pointer is taken to be a Gaussian centered around zero, and the canonical conjugate P is bounded and also a motion constant of both the interaction Hamiltonian and the whole Hamiltonian.

According to the principle of protective measurement, the measurement of A_n yields the following result:

$$\langle A_n \rangle = \frac{1}{v_n} \int_{v_n} |\psi(x)|^2 dv = |\psi_n|^2 \tag{4.8}$$

It can be seen that the result $\langle A_n \rangle = |\psi_n|^2$ is the average of the position measure density $\rho(x) = |\psi(x)|^2$ over the small region V_n. Then when $v_n \to 0$ and after performing measurements in sufficiently many regions V_n we can find the position measure density $\rho(x)$ of the quantum motion of the particle.

In order to find the position measure current density $j(x)$, we need to measure the value of an observable $B_n = \frac{1}{2i}(A_n \nabla + \nabla A_n)$. According to the principle of protective measurement, the measurement result is:

$$\langle B_n \rangle = \frac{1}{v_n} \int_{v_n} \frac{1}{2i}(\psi^* \nabla \psi - \psi \nabla \psi^*) dv = \frac{1}{v_n} \int_{v_n} j(x) dv \tag{4.9}$$

It can be seen that the result is the average value of the position measure flux density $j(x)$ in the region V_n. Then when $v_n \to 0$ and after performing measurements in sufficiently many regions V_n, we can also find the position measure flux density $j(x)$ of the quantum motion of the particle.

In a word, we have shown that the quantum motion of a particle, which is described by the position measure density $\rho(x)$ and the position measure flux density $j(x)$, can be more directly revealed by the above protective measurement. It should be stressed that a small ensemble of similar particles may be needed for protective measurement in real experiments. In addition, we can complete the measurement of charged particles more easily, for which $\rho(x,t)$ and $j(x,t)$ represent the effective charge density and current density.

As an example, we briefly analyze the well-known double-slit experiment. As we know, taking a typical position measurement near the slits will destroy the double-slit interference pattern. This kind of measurement cannot reveal the objective motion state of the particle passing through the two slits. By comparison, protective measurement may help to reveal the quantum motion of the particle passing through the two slits. According to the principle of protective measurement, given that we know the state of the particle beforehand in the double-slit experiment, we can protectively reveal the objective motion state of the particle when it passes through the two slits. At the same time, the motion state of the particle will not be destroyed after a protective measurement, and the interference pattern will not be destroyed either. As we have shown, the results of the protective measurement will show that the position measure density of the particle is distributed throughout both slits. For example, in the double-slit experiment of electrons, the protective measurement will show that there is a charge of $e/2$ in each of the two slits when the single electron is passing the slits. Since the measurement result is in principle irrelevant to the duration of the measurement, there will be a charge of $e/2$ in each of the two slits during an arbitrarily short time interval within the precision of measurement. This result clearly shows that the single particle passes through both slits in the double-slit experiment, and its motion is indeed discontinuous.

CHAPTER 5

Understanding Quantum Motion

Quantum motion needs to be understood. If we don't understand quantum motion, we cannot understand motion at all. Quantum motion needs to be understood. We live in the macroscopic world, and we are only familiar with continuous motion. But continuous motion is only the approximate display of quantum motion in the macroscopic world. Quantum motion needs to be understood. We have entered into the microscopic world, but nobody knows what is happening there. We have a theory named quantum mechanics, but no one understands it. The quantum conundrum, which had puzzled us for a century, still puzzles us today. Quantum motion may be the answer.

For the convenience of discussions, we will mainly analyze the random discontinuous motion (RDM) in continuous space-time in this chapter. Most of the analyses also hold true for quantum motion, which is defined as the RDM in discrete space-time. In addition, in order to understand quantum motion more easily, we often compare quantum motion with continuous motion in the discussions. This may be an easier way to understand quantum motion. However, we stress that continuous motion is not an actual form of motion, but only the ideal approximation of the real quantum motion in the macroscopic world.

5.1 What on Earth Does the Wave Function Tell Us?

Since the wave function ψ was found by Schrödinger in Arosa, people have been disputing with each other on its meaning. What on earth does ψ tell us? In this section we will show that if ψ is a complete description of the motion of particles, then it has told us that what it describes is the RDM of particles.

As we know, the motion state of a particle is defined in an infinitesimal time interval. If the wave function $\psi(x,t)$ provides a complete description of the motion state of a particle, the usual probability distribution $|\psi(x,t)|^2$ will represent the real position distribution of the particle in an infinitesimal space interval near position x during an infinitesimal time interval near instant t. This means that the particle will move throughout the space where the wave function $\psi(x,t)$ spreads during an infinitesimal time interval, although the particle is still in one position at each instant. This kind of motion is essentially discontinuous.

Whereas the wave function ψ is a universal description of the motion of particles, the motion of particles will be universally discontinuous. The properties of the particle undergoing RDM will not only include position, but also include all the other properties such as momentum and spin etc, which can be in a quantum superposition. A detailed analysis of RDM has been given in the preceding chapters. Such an analysis has demonstrated that ψ is the very mathematical complex describing the RDM of particles, and the Schrödinger equation is also its simplest nonrelativistic evolution equation.

This is what ψ really tells us. Its voice is so melting and real!

5.2 How Does An Object Move From *A* To *B*?

As to the weird displays of RDM, people may naturally ask a question, i.e., that how a particle moves from position *A* to position *B* in space.

The answer seems to be very simple for continuous motion. It is that the particle moves from position *A* to position *B* along a continuous trajectory in space. But after a careful analysis, we will find that where the particle is at each instant during its movement from position *A* to position *B* must finally be examined before the answer of continuous trajectory can be given. Similarly, as to RDM, we still need to examine where the particle is at each instant during its movement from position A to position B. However, the answer is not a continuous trajectory, but a discontinuous point set. Concretely speaking, the particle undergoing RDM is still in one position at each instant during its movement from position A to position B. But these positions do not form a continuous trajectory, but form a random discontinuous point set, in which the positions at adjacent instants are generally not adjacent. This is the answer of RDM to the above question.

However, most people probably have a dissatisfactory feeling about the answer of RDM. The dissatisfaction may result from the following fact: the answer of the above question can lead to a law for continuous motion, but it cannot lead to any law for RDM. In the following, we will further analyze the causes resulting in the dissatisfaction.

When studying the motion of objects in the macroscopic world, people have been accustomed to ask the above question about the trajectory of an object, and acclaim the great achievements of classical mechanics for its accurate answer to this question. It is indeed this question that leads to the naissance of modern science to some extent. However, relating the answer of this question to the motion law may result in a prejudice, namely regarding the instantaneous state of a particle as the motion state of the particle. In fact, the motion state of a particle should refer to the state of the particle during an infinitesimal time interval. The instantaneous state of a particle only contains the existence of the particle, and cannot reflect the motion of the particle at all. But the common sense about the continuity of motion has been preventing people from finding the essential limitations of the instantaneous definition of the motion state of a particle. Indeed, as to continuous motion, the instantaneous definition of the motion state of a particle is a reasonable simplification of the interval definition of the motion state of a particle in both mathematical description and physical meaning. In mathematics, the differential evolution equations of continuous motion, which are based on infinitesimal analysis, generally possess explicit function solutions containing instantaneous variables. In physics, the continuity of continuous motion guarantees that the instantaneous description and the interval description are equivalent in both theory and experiment. These facts make people be more inclined to emphasize the explicit function solutions in the understanding of motion, as well as in the application of classical mechanics, and disregard the original differential description of motion, which refers to infinitesimal intervals and possesses real physical meaning. In a word, the instantaneous definition of the motion state of a particle is valid and applicable in the domain of continuous motion. This strengthens people's belief in it.

However, as to RDM, the instantaneous definition of the motion state of a particle will be essentially improper. The reason is evident. Since the continuity precondition, which leads to the reasonable simplification of this definition, disappears for RDM, and is replaced by the essential discontinuity. This kind of discontinuity thoroughly reveals the inherent irrationality of the instantaneous definition of the motion state of a particle, and makes people renewedly realize the facticity and fundamentality of the interval definition of the motion state of a particle. As a result, the above question about the trajectory of a particle is not helpful for finding the law of RDM. For RDM, the trajectory of a particle is discontinuous and random everywhere, and no law exists for the trajectory of the particle. Accordingly people's dissatisfaction with the above answer of RDM is out of all reason.

5.3 Instant and Infinitesimal Time Interval

In this section, we will further analyze the essential differences between instant and infinitesimal time interval. Some of these differences are as follows.

(1). The instantaneous state of a particle contains only one point in space, while the infinitesimal interval state of a particle contains uncountable points in space. In mathematics, the cardinal number of a set containing one element is zero, while the cardinal number of a set containing uncountable elements is larger than $\aleph_0$, the cardinal number of the natural number set.

(2). The instantaneous state of a particle contains no motion, but only the existence of the particle. The infinitesimal interval state of a particle may contain abundant elements of motion, since it contains uncountable points in space.

(3). The instantaneous state of a particle possesses no physical meaning, since we cannot measure it in physics. The infinitesimal interval state of a particle possesses physical meaning, since we can measure it through the process $\Delta t \rightarrow dt$ in principle.

(4). We can only test the evolution law of the infinitesimal interval state of a particle. Even if the evolution law of the instantaneous state of a particle exists, we cannot test it.

Since the state of a particle at one instant contains no motion, the instantaneous state of a particle is not the motion state of the particle in any case. We cannot conclude that the motion state of a particle must be a local state in space either. The motion state of a particle should relate to the state of the particle during a time interval, and can be defined as the state of the particle during an infinitesimal time interval in mathematics. As to the position state of a particle during an infinitesimal time interval *dt*, there exists no *a priori* reason to require that it must take some kind of special form, say a continuous line localized in an infinitesimal space interval *dx*. On the contrary, a natural assumption in logic is that the position state of the particle during an infinitesimal time interval is a random discontinuous point set in which the points spread over the whole space. This is just the motion state of the RDM of a particle. Thus RDM naturally accords with the interval definition of the motion state of a particle, although it looks very bizarre and unnatural according to the improper instant definition of the motion state of a particle. Due to the adoption of the instant definition of motion state, people have been discussing the motion of particles in the framework of point. But from both physical and mathematical considerations, motion should be studied in the framework of point set. In the framework of point set, continuous motion looks very bizarre and unnatural according to the proper interval definition of the motion state of a particle.

In the following, we will further analyze the differences between instant and infinitesimal time interval in the descriptions of continuous motion and RDM. As to continuous motion, the instantaneous description and the infinitesimal interval description are equivalent. The former will result in the intuitionistic trajectory description of motion such as $x = x(t)$, while the latter will result in the strict differential description of motion such as $dx/dt = v(x,t)$. Although the differential evolution equations of continuous motion have no explicit function solutions containing instantaneous variables for some situations, the instantaneous state and the infinitesimal interval state can both be approached with arbitrary precision in experiment from the motion state during a finite time interval due to the essential continuity of continuous motion. Thus the instantaneous description and the infinitesimal interval description are equivalent for continuous motion in physics. Their difference only lies in metaphysical meaning. The former favors the opinion that matter and motion are separable, while the latter favors the opinion that matter and motion are inseparable.

As to RDM, the instantaneous description and the infinitesimal interval description are essentially different. Although there still exists an instantaneous description such as $x = x(t)$ for RDM, such description no longer possesses any physical meaning owing to the essential discontinuity of RDM, and there exists no law for the instantaneous state either. For RDM, the instantaneous state cannot be approached from the motion state during a finite time interval in experiment. These two kinds of states are essentially different. One is the local existence of a particle, and the other is the non-local existence of the motion of a particle. By comparison, the infinitesimal interval state of RDM can be approached with arbitrary precision from the motion state during a finite time interval. Thus the infinitesimal interval description still possesses physical meaning for RDM, and there also exists an evolution law for the infinitesimal interval state. Since the infinitesimal interval state of RDM is different from that of continuous motion, the explicit function solutions of the evolution equation of RDM are no longer instantaneous position functions such as $x(t)$, but new interval descriptive functions such as the position measure density $\rho(x,t)$.

5.4 Velocity and Momentum

For RDM, the meanings of velocity and momentum are very different from those in classical mechanics. We need to understand them in order to understand RDM.

An evident character of RDM is that the particle undergoing RDM does not possess velocity. In fact, the particle discontinuously appears in different positions of space, and it has no continuous trajectory at all. An intuitional picture of RDM is that it spreads in space like a cloud. This cloud-like stuff is generated by the whole discontinuous position set of a particle during an infinitesimal time interval. The stuff is denser in the region where the position measure density of the particle is larger, and is sparser in the region where the position measure density of the particle is smaller. Such stuff is often called wave packet in the textbooks of quantum mechanics. The spreading of the cloud-like stuff can be described by an average velocity, which is defined as the group velocity of the corresponding wave packet. Thus although there exists no strict velocity description similar to that of continuous motion, we can still have a rough description using average velocity for RDM.

In addition, there always exists a rest state for the continuous motion of a particle when selecting a suitable frame of reference. This facilitates the study of the particle in motion. But as to the RDM of a particle, since the particle discontinuously moves at all times, there exists no rest state for the particle in essence. However, there is still a rest state for the RDM of a particle. It is defined as the motion state of the particle which does not change with time.

Concretely speaking, the position measure density $\rho(x,t)$ and the position measure flux density $j(x,t)$ do not change with time in the rest state. Such state is often called stationary state. The intuitional picture of stationary state is that the cloud-like stuff generated by the RDM of the particle is at rest in space. It should be noted that since the cloud-like stuff or wave packet tends to spread to a larger region, a stationary state needs to be bounded by an external potential. This is essentially different from the situation in classical mechanics.

Undoubtedly, the above intuitional pictures based on the familiar velocity concept make RDM be more intelligible. However, in order to understand RDM, we must also face the abstract property momentum.

First, momentum is an intrinsic property of the RDM of a particle, and is as basic as position for the description of motion. The momentum of a particle is no longer defined as its velocity multiplied by its mass, since there exists no velocity for a particle undergoing RDM. However, the momentum of a particle relates to the velocity of the whole discontinuous position set, which is formed by the RDM of the particle during an infinitesimal time interval, to some extent[12]. Especially, the average momentum of a particle directly relates to the velocity of the whole discontinuous position set. Thus we can say that momentum determines the motion of a particle in the meaning of time average, in other words, the momentum of a particle determines its propensity to move at each instant. Although the motion of a particle is random, there still exist some properties such as momentum which determines its propensity to move. This then generates the lawful continuous evolution of the whole discontinuous position set. As a result, there should exist a one-to-one relation between the position

[12] Note that the momentum of a particle does not relate to the velocity of the continuous movement of the local discontinuous position set. It is the position measure flux density $j(x,t)$ that relates to such a velocity.

description and the momentum description. This is consistent with the preceding analysis in Section 2.3.

Next, momentum is a non-local property. For a free particle with a constant momentum, its position will not be limited in an infinitesimal space interval dx during an infinitesimal time interval dt, but spread throughout the whole space with the same position measure density $\rho(x,t)=1$. This clearly shows that momentum possesses one kind of non-local character. By comparison, position is evidently a local property.

Thirdly, just like the motion picture in position space or real space, there also exists a similar motion picture in momentum space. The momentum of a particle assumes a definite value at each instant, but it randomly spreads through all possible values with a certain momentum measure density $f(p,t)$ during an infinitesimal time interval. These two kinds of motion pictures are related through the one-to-one relation $\psi(x,t) = \int_{-\infty}^{+\infty} \varphi(p,t)e^{ipx-iEt}dp$ between the position description and the momentum description.

Fourthly, the relation between the position and the momentum of a particle are essentially discontinuous and random at all instants, i.e., there exists no a continuous function $p(x,t)$ or $x(p,t)$. Accordingly there exists no any nontrivial joint distribution function of position and momentum for the RDM of a particle. This is also a result of quantum mechanics. In addition, it should be stressed that the fact that a particle possesses a definite position and a definite momentum at each instant does not contradict Heisenberg's uncertainty principle. According to the one-to-one relation between the position description and the momentum description, the standard deviations of the distributions of position and momentum always satisfy Heisenberg's uncertainty relation for any motion state of a particle. This is a strict mathematical result. Moreover, due to the happening of the wavefunction collapse during measurement, we cannot precisely measure the position and momentum of a particle simultaneously either.

Lastly, we can naturally generalize the RDM from real space to other abstract spaces. Then the motion picture of a particle will be that at any instant the properties of the particle such as energy and spin etc all have definite values, and during an infinitesimal time interval they spread through all possible values with respective measure density. In general, the correlation between the instantaneous values of different physical quantities is discontinuous and random.

5.5 Quantum Motion and Copenhagen Interpretation

Copenhagen interpretation is the widely accepted interpretation of quantum mechanics. It has many reasonable elements. Thus studying the relation between the theory of quantum motion and Copenhagen interpretation will undoubtedly be helpful for the understanding of quantum motion. In addition, it will also make people grasp Copenhagen interpretation more easily. Here we will give a detailed analysis of their relations.

According to Primas (1981)'s summary, Copenhagen interpretation can be outlined as follows.

(1) The theory is concerned with individual objects.

(2) Probabilities are primary.

(3) The frontier separating the observed object and the means of observation is left to the choice of the observer.

(4) The observational means must be described in terms of classical physics.

(5) The act of observation is irreversible and it creates a document.

(6) The quantum jump taking place when a measurement is made is a transition from potentiality to actuality.

(7) Complementarity properties cannot be observed simultaneously.

(8) Only the results of a measurement can be taken to be true.

(9) Pure quantum states are objective but not real.

In the following, we will analyze the above summary of Copenhagen interpretation one by one. The first item, i.e., that quantum mechanics is concerned with individual objects, clearly states that quantum mechanics is not a theory about the ensemble comprising a large number of particles, but a theory about individual objects. Thus Copenhagen interpretation excludes the possibility of an ensemble interpretation of quantum mechanics. This is consistent with the theory of quantum motion, according to which what quantum mechanics describes is the RDM of individual objects. Furthermore, the theory of quantum motion also provides a convincing exemplification for the first item of Copenhagen interpretation.

Concerning the second item, i.e., that probabilities are primary, it means that the probabilities appearing in quantum mechanics don't result from the ignorance of the observer or the inability of the theory, but must be regarded as an essential character of Nature. Moreover, when quantum mechanics is able to predict these probabilities, it should be accepted as a complete theory. No doubt it is extremely difficult or even impossible to understand this item within Copenhagen interpretation. As a result, people always resort to the law of causation to reject this item, and further get back the classical deterministic picture by introducing the so-called hidden variables. Then they naturally regard the appearance of probabilities in quantum mechanics as an indication of the incompleteness of the theory. Which of these two viewpoints is correct cannot be determined before we find the real motion in quantum world. In fact, it has been a vexed problem since the founding of quantum mechanics. Now in the light of the theory of quantum motion, the probabilities appearing in quantum mechanics result from the actual quantum motion of particles, which is essentially discontinuous and random. Accordingly the theory of quantum motion provides a real physical explanation of the second item of Copenhagen interpretation.

The third item, i.e., that the frontier separating the observed object and the means of observation is left to the choice of the observer, is evidently unsatisfactory. It does not give a quantitative physical description to determine the border and further distinguish between the observed object and the measuring apparatus. Although Bohr evaded this difficulty by regarding the observed object and the measuring apparatus as an indivisible whole, his opinion may be inconsistent. The macroscopic measuring apparatus is regarded as one kind of independent existence, while a macroscopic measuring apparatus is composed of a large number of microscopic particles, thus it must be an *ad hoc* prescription not to regard the microscopic particles as one kind of independent existence. In fact, there should exist an objective physical border between the microscopic particles and the macroscopic measuring apparatus, and we must give an accurate quantitative description of this border. However, such description is missing in quantum mechanics and its Copenhagen interpretation. Now the theory of quantum motion provides a quantitative physical description of the border, and objectively explains the wavefunction collapse resulting from the interaction between the microscopic particles and the macroscopic measuring apparatus. Especially it gives a uniform realistic description of the microscopic and macroscopic worlds. Thus the theory of quantum motion is more satisfactory than Copenhagen interpretation.

The fourth item, i.e., that the observational means must be described in terms of classical physics, means that although the observed microscopic objects are so peculiar that classical physics can no longer provide a consistent explanation of their displays, our observational means must still be described in terms of classical physics, and we can only use classical concepts to describe the experimental facts. This conclusion has been widely accepted. But as we think, all concepts are only free inventions of human being, and their validity and applicability must be verified at any moment. Especially when there appears a new experimental fact that cannot be consistently explained by the existing concepts, this kind of examination is more indispensable. Thus we must examine and validate the validity and completeness of the classical concepts. On the other hand, even if we still use the existing classical concepts, we must be ready to renewedly understand their meanings in face of new experience. The fact is always that we don't really understand the concepts invented by us in the beginning, and this kind of understanding becomes deeper and deeper only when more and more experience is accumulated. Now the universal existence of quantum motion makes people see the limitations of classical concepts more clearly. Even for the observational means and the macroscopic phenomena they are only approximate descriptions, and their existence cannot prevent us from finding the description which is closer to reality.

The fifth item, i.e., that the act of observation is irreversible and it creates a document, holds true for both the observation on the microscopic systems and that on the macroscopic systems. Moreover, the irreversible process can be explained only in terms of classical physics. Thus the irreversible process during an observation is actually irrelevant to the peculiar properties of quantum measurement. Now quantum motion and its law further confirm this conclusion.

The sixth item, i.e., that the quantum jump taking place when a measurement is made is a transition from potentiality to actuality, explicitly asserts the objective existence of instantaneous wavefunction collapse during a measurement. Furthermore, Copenhagen interpretation acknowledges that the quantum jump or instantaneous wavefunction collapse is a new physical process, and cannot be accounted for by quantum mechanics. As we think, it is just the absence of the description of this process that results in the physical incompleteness of the existing quantum theory. Owing to this absence, Copenhagen interpretation is not a complete interpretation either. Now the theory of quantum motion provides an objective description of the wavefunction collapse process. It is a complete quantum theory, and naturally includes a complete interpretation.

The seventh item, i.e., that complementarity properties cannot be observed simultaneously, is the core of Copenhagen interpretation (cf. Bohr 1927). But as we think, it is also the most obscure part of the interpretation. Although this assertion is correct, its demonstration given by Copenhagen interpretation is by no means complete. On the one hand, Copenhagen interpretation regards this assertion as an inevitable result of the measurement disturbance. However, it never provides a clear explanation of this disturbance, and its demonstration is always a mixture of classical part and quantum part. In fact, in order to understand the measurement process, we must deal with the quantum entanglement process and the wavefunction collapse process during the measurement. On the other hand, Copenhagen interpretation overemphasizes the influence of the measurement disturbance, and disregards the possibility that the objective motion state of the observed object may be the main physical cause. The fact that complementarity properties cannot be observed simultaneously may actually reflect the peculiarity of the motion state of the observed object. Since Copenhagen interpretation denies the existence of the objective motion of microscopic particles, its demonstration of the above assertion cannot be complete.

Now the theory of quantum motion will provide a clear and complete physical explanation for the above assertion. Here as an example we discuss the observation on the complementarity properties position and momentum. First, concerning any quantum motion state of a particle, the standard deviations of the distributions of position and momentum all satisfy Heisenberg's uncertainty relation. Concretely speaking, as to the motion state in which the position of the particle is more definite, the momentum distribution of the particle will be closer to an even distribution, i.e., the momentum of the particle will be more indefinite, or vice versa. Then there exists no a motion state in which the position and the momentum of the particle are both definite in reality. Moreover, according to the reasonable assumption that measurement truly reflects the observed state, this kind of quantum motion state will further require that the position and the momentum of a particle cannot be observed simultaneously. It should be noted that, whereas the position and the momentum of a particle cannot both be in a definite state in reality, it may be improper to say that the position and the momentum of a particle cannot be observed simultaneously, since this statement seems to imply that the position and the momentum of the particle can both be in a definite state before measurement. Secondly, when considering the measurement process, the position measurement of a particle will result in the wavefunction collapse process, and the motion state of the particle will collapse to one of its position eigenstates, in which the momentum distribution of the particle is an even distribution, according to the law of quantum motion. This conclusion is the same for the momentum measurement of a particle. Thus the above assertion given by Copenhagen interpretation can be physically explained by the theory of quantum motion.

As to the last two items, i.e., that only the results of a measurement can be taken to be true, and pure quantum states are objective but not real, they mean that there exists no any realistic picture for the microscopic objects. This is an astonishing assertion. In the following, we will give a critical analysis of this assertion[13].

As we know, Bohr repeatedly stressed that any elucidation of the microscopic phenomena must resort to complementarity principle. Concretely speaking, the information obtained by the measurements under different experimental conditions will exhaust all definable knowledge about the observed microscopic object, but at the same time, when we try to unite the information in a realistic picture it appears to be incompatible. Then any single realistic picture cannot provide an exhaustive account of the microscopic phenomena, instead we can only provide a complementary account of the microscopic phenomena using incompatible classical pictures. Accordingly Copenhagen interpretation asserts that there exists no any realistic picture for the microscopic objects, and we can only use the complementary classical pictures to describe them.

---

[13] It is worth noting that (d'Espagnat 2003, 2006) convincingly argued that the notion of an ultimate reality is conceptually necessary, although such a reality may be a "veiled reality".

It can be seen that the essential reason why Copenhagen interpretation rejects the realistic picture of the microscopic objects is that the information obtained by the measurements under different experimental conditions is incompatible when being united in a single realistic picture. Then why is the information incompatible when being united in a single realistic picture? Which picture is the realistic picture in which the information is incompatible? The answer of Copenhagen interpretation is that the information is incompatible when being united in a single picture of classical corpuscular or classical wave. The information obtained by the measurements under some experimental conditions shows that the display of a microscopic object resembles that of classical corpuscular, while the information obtained by the measurements under other experimental conditions shows that the display of a microscopic object resembles that of classical wave. Thus the realistic pictures rejected by Copenhagen interpretation are simply the pictures of classical corpuscular and classical wave. Then are there any further reasons to reject all the other possible realistic pictures? No! If there is one reason, it is only the prejudice unconsciously hold by most people including Bohr and Einstein, i.e., that the pictures of classical corpuscular and classical wave are the only possible realistic pictures, in other words, continuous motion is the only possible form of motion.

It should be acknowledged that the pictures of classical corpuscular and classical wave are indeed helpful for describing the microscopic objects. But how can we prove that this kind of description is the only possible description? And why must we still use the macroscopic classical pictures to describe the microscopic processes? The picture of continuous motion directly comes of our macroscopic experience, and the theory of continuous motion, namely classical mechanics, had gained some successes. But as we know, continuous motion is no longer suitable for describing the microscopic objects, then whereby do we think continuous motion is the only possible motion? And whereby do we assert that the single realistic picture of the microscopic objects does not exist?

In fact, the appearance of quantum mechanics has strongly implied that there exists a new kind of motion which is different from the familiar continuous motion. The new motion will provide a single realistic picture for the microscopic objects, and can naturally display the complementarity properties which are mutually exclusive in the framework of classical physics. Quantum mechanics does not prevent us from finding the new motion. What block us are only ourselves, our prejudice, our arrogance and our ignorance. Now the existence of quantum motion clearly reveals the limitations of complementarity principle. At the same time, it makes people have to painfully reject the prejudice of the uniqueness of continuous motion. The pain is transitory, while the happiness brought by the understanding of reality is permanent.

In a word, owing to the absence of a realistic picture of the microscopic objects, the above essentials of Copenhagen interpretation can hardly be convincing, and the intrinsic connections between them can hardly be established either. Moreover, complementarity principle, which is the core of Copenhagen interpretation, also proves to be trustless. By comparison, the theory of quantum motion provides a real picture of microscopic reality, and can establish the intrinsic connections between the valid essentials of Copenhagen interpretation. But accordingly Copenhagen interpretation no longer exists, and will be replaced by the new realistic interpretation in terms of quantum motion.

5.6 Understanding the Displays of Quantum Motion

In this section, we will analyze several well-known displays of quantum motion in detail. These displays will help us understand quantum motion more easily.

5.6.1 The Stability of the Hydrogen Atom

The stability of the hydrogen atom is a big puzzle in the beginning of the 20th century. The essential reason lies in that classical physics of the day is incapable of accounting for it. However, it is still a great puzzle even today. Even if, as most people think, quantum mechanics has solved it, but, as everyone consents, no one today understands quantum mechanics.

In order to solve the knotty stability problem, Bohr first presented three well-known assumptions in 1913. Although these assumptions established some connections in the experimental data of the spectra of the hydrogen atom, they didn't explain the stability of the hydrogen atom. On the contrary, Bohr simply assumed the stability of the hydrogen atom. In the 1920s, quantum mechanics was founded. By applying the theory people can readily calculate the energy of the hydrogen atom in all stationary states, and the predictions accurately accord with the experimental data. However, quantum mechanics does not account for the stability of hydrogen atom either. It just uses one mystery, the wave function and its equation, to explain another mystery, the stability of the hydrogen atom. The real physical origin of the stability has not been found yet. To our surprise, owing to the appearance of quantum mechanics, especially its striking successes in accounting for the microscopic phenomena, the stability problem faded from people's memories as time goes on. Most people today take it for granted that this problem has been completely solved by quantum mechanics. It is a pity for science. It is also a joke for human reason.

Now we will give the physical origin of the atomic stability in terms of the theory of quantum motion. As a simple example, we analyze the base state of the Hydrogen atom, whose position measure density is:

$$\rho(r,\theta,\varphi)=|\psi(r,\theta,\varphi)|^2=\frac{4}{a_0}\exp(-\frac{2r}{a_0}) \tag{5.1}$$

where a_0 is the Bohr radius. According to the picture of quantum motion of a single particle, the electron will still be in one position at each instant, but during an infinitesimal time interval dt the electron will move throughout the whole space with the above position measure density. Since the position measure density of the electron possesses spherical symmetry, the charge distribution of the electron will be equivalent to a negative elementary charge in the center of sphere. This just counteracts the positive elementary charge of the nucleus in the center of sphere. Then the charge distribution of the whole system including electron and nucleus will be equivalent to a zero charge distribution in physics. In addition, since the electron is in the base state, an energy eigenstate, the stochastic term in the evolution equation of quantum motion disappears. Then the position measure density of the electron will not be changed during the evolution, and accordingly the zero charge distribution of the whole system will not change either. Thus no electromagnetic energy is radiated for the system, and the base state of the hydrogen atom is stable.

In a word, the theory of quantum motion provides a physical explanation of the mysterious stability in the atomic world. It is just this kind of stability that makes the macroscopic world be stable, and further permits human beings to appear to understand it.

5.6.2 Double-slit Experiment

Double-slit experiment has been widely discussed since the founding of quantum mechanics. Nearly all textbooks of quantum mechanics illustrate the weirdness of quantum world using the experiment. As Feynman (1963) said, it contains the only mystery of quantum mechanics. But have we disclosed the mystery and understood the double-slit experiment? As we think, the answer is definitely negative. The realists have been asking, "how on earth does the single particle pass through the two slits?[14]" In this section, we will present a clear realistic picture of the particle passing through the two slits to form the double-slit interference pattern, and demonstrate that the double-slit interference is just the natural display of quantum motion.

The right figure is a simple sketch map of double-slit experiment. The single particle (e.g. electrons) is emitted from the source S one after the other, then passes through the two slits to arrive at the screen B. In this way, when a large number of particles arrive at the screen, they collectively form a double-slit interference pattern.

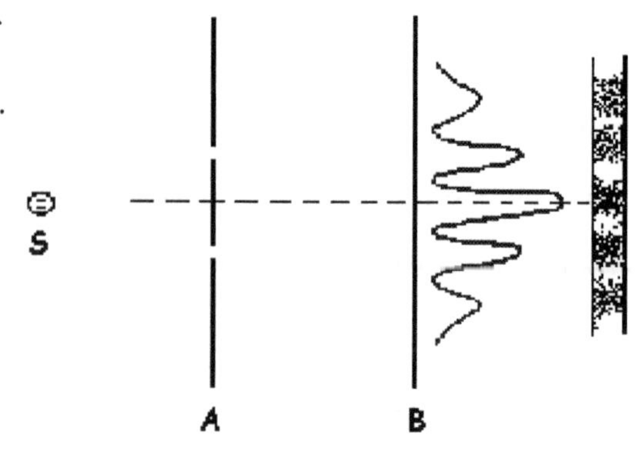

Figure 6 Double-slit experiment

Let's first see whether or not the familiar continuous motion of particles can account for the formation of the double-slit interference pattern. According to the definition of continuous motion, the single particle can only pass through one of the two slits in each experiment. Then it is evident that the double-slit interference pattern will be the same as the direct mixture of two one-slit patterns, each of which is formed by opening each of the two slits. The passing process of each particle in double-slit experiment is exactly the same as that in one of these two one-slit experiments. However, all known experiments show that the interference patterns for the above two situations are evidently different. This is an inevitable dilemma when using continuous motion to explain the formation of the double-slit interference pattern. In fact, we can see where the perplexity lies more easily from the following fact, i.e., that when one of the two slits is shut, the particle can reach some position on the screen, but when the shut slit is opened, it will prevent the particle from reaching the above position on the screen.

---

[14] Indeed, it is just this question that touches our sore spots in understanding quantum mechanics, and unveils the deadly flaw of the orthodox interpretation of quantum mechanics.

No way out but reject continuous motion. The orthodox interpretation of quantum mechanics indeed rejects continuous motion, but at the same time, it also rejects all possible forms of motion, and proves that the rejection is inevitable. Thus the orthodox interpretation not only fails to give the realistic motion picture of the particle passing through the two slits, but also surprisingly asserts that this is not due to its inability, but because the motion picture does not exist at all in reality. In the following, we will see how the orthodox interpretation "practises deception", and where it "gives the show away".

The orthodox interpretation first implicitly assumes that continuous motion is the only possible form of motion, and then proves that continuous motion cannot account for the double-slit interference pattern by employing the similar demonstration as the above. Thus the orthodox interpretation rejects continuous motion, and owing to the uniqueness of continuous motion, it also rejects all possible forms of motion for the microscopic particles. It asserts that when people talk about any property of particles, they must measure the property, in other words, there exists no microscopic reality independent of observation. Furthermore, the orthodox interpretation explains the double-slit experiment in the sense of measurement, and regards such an explanation as the only possible one. This explanation can be simply expressed as follows. If you want to know how the single particle passes through the two slits to form the double-slit interference pattern, you must detect which slit the particle passes through by taking a position measurement. But according to quantum mechanics, this kind of position measurement will inevitably destroy the double-slit interference pattern. Then on condition that the double-slit interference pattern is not influenced, we cannot detect which slit the single particle passes through, and thus we cannot know how the single particle passes through the two slits to form the double-slit interference pattern. Accordingly the realistic motion picture of the particle passing through the two slits does not exist in the sense of observation. Moreover, since there exists no microscopic reality independent of observation, the realistic motion picture of the particle passing through the two slits does not exist in essence.

The above demonstration of the orthodox interpretation seems to be flawless, and it indeed "dusts the eyes of" nearly all great men in the 20th century[15]. However, there exist two unnoticed deadly flaws in the demonstration. One is that the orthodox interpretation implicitly assumes continuous motion is the only possible form of motion, but it never gives a justifiable argument. The other is that the orthodox interpretation only employs the position measurement to detect how the particle passes through the two slits in individual experiment.

Concerning the first flaw, the implicit assumption of the uniqueness of continuous motion has never been seriously examined and disbelieved. Even we can say, no one has ever noticed that it is an assumption, since nearly all people including those arguing against the orthodox interpretation religiously believe in it. The validity of this assumption appears to be evident, but as we will see, it is an ingrained prejudice. It was fed to grow up by the successful experience and the edification of great men, but it instead fetters the thoughts of the great men, and tries to obliterate the reality behind experience. It finally becomes a terrible demon under the cosher of people, and even wants to devour the whole real world. Now only reason can tame it, and help it get back the lily-white heart and the naive temperament. Indeed, there exist many reasons to lead people to unconditionally accept the above assumption, and the reasons from experience and history may play a decisive role. But we must note that people seldom consider the rationality of this assumption, and never seriously study whether there exist other possible even more fundamental forms of motion, even though they cannot help but reject continuous motion in the face of quantum mechanics. Why do people believe in it so religiously? An interesting fact may be that it is unnecessary for people to doubt this

[15] The underlying reason, as we think, is that they all had the following prejudice, i.e., that continuous motion is the only possible form of motion.

assumption before the appearance of quantum mechanics, while the orthodox interpretation prohibits people from doubting it again after the founding of quantum mechanics. But then, we had better leave this question to the philosophers.

Now we will further analyze the first flaw in the demonstration of the orthodox interpretation. As we know, when discussing the double-slit experiment, people always ask the following question, i.e., that which slit the single particle passes through in individual experiment. The orthodox view asserts that this question is meaningless, since we cannot measure which slit the particle passes through on condition that the interference pattern is not destroyed. In fact, this question is indeed meaningless, and as it happens the orthodox answer is right. But as we have seen above, its reason is by no means right. The actual reason should be that if the particle passes through only one slit in each experiment, the interference pattern will not be formed at all[16]. Thus it is obviously wrong to ask which slit the single particle passes through in individual experiment. The particle must pass through more than one slit, otherwise the double-slit interference pattern cannot be formed.

On the other hand, we can still ask the following question, i.e., that how the single particle passes through the two slits to form the interference pattern. It is just this question that clearly unveils the first deadly flaw of the orthodox interpretation. Then what is the answer of the orthodox interpretation? As we know, it concludes that there exists no any realistic motion picture of the particle passing through the two slits, and the above question is still meaningless. But how can it arrive at this conclusion? It cannot do, and no one can do. What the orthodox interpretation bases on is only the above prejudice, i.e., that continuous motion is the only possible form of motion. Once the prejudice is rejected, and the realistic motion picture of the particle may exist, we can actually find it by using a logical microscope. The reasoning is very

[16] Here we assume the particle is the only existence as the orthodox interpretation does. Thus Bohm (1952)'s hidden-variable theory is not considered.

simple. Since the particle does not pass through only one slit in individual double-slit experiment, it must pass through both slits when passing through the two slits. The particle has no other choices in logic! This kind of bizarre motion is not impossible in principle. Since it will take a period of time, not one instant, for the particle to pass through the slits, what the particle needs to do is just moving discontinuously in the two slits. In reality, this is just the display of quantum motion.

Now we will turn to the second flaw in the demonstration of the orthodox interpretation, i.e., that the orthodox interpretation only uses the position measurement to detect how the particle passes through the two slits in individual experiment. This flaw is a technical flaw, and actually results from the first flaw. The validity of position measurement relies on the absolute validity of the implicit assumption in the first flaw, according to which continuous motion is the only possible form of motion. When this assumption is no longer valid, and the realistic motion of particles does exist and is not continuous motion, the single particle will pass through both slits in a discontinuous way. Then the position measurement may be incapable to detect how the particle passes through both slits in individual experiment, and thus it may be invalid. The orthodox interpretation always pertinaciously takes the position measurement, but it just swallows the bait of quantum mechanics. Quantum mechanics can easily meet such position measurement by using the wavefunction collapse, and thus it successfully conceals the real face of the microscopic reality. According to quantum mechanics, the position measurement will destroy the real motion state of the particle, and collapse it to the region near one slit. As a result, the position measurement not only destroys the double-slit interference pattern, but also is incapable to detect the real motion picture of the particle passing through the two slits.

Once we realize the above technical flaw in the demonstration of orthodox interpretation, we can try to use a new kind of measurement. It can meet the situation where the particle passes through both slits in individual experiment, and can detect the real motion state of the particle passing through the two slits while not destroying the interference pattern. Fortunately, this kind of measurement has been found (cf. Aharonov and Vaidman 1993; Aharonov, Anandan and Vaidman 1993). It is called protective measurement. According to the theory of protective measurement[17], since we know the wave function of the particle beforehand in double-slit experiment, we can protectively measure the real motion state of the particle when it passes through the two slits. At the same time, the wave function of the particle will not be destroyed after such protective measurement, and the interference pattern will not be destroyed either. Thus by using protective measurement we can find the realistic motion picture of the particle passing through the two slits while not destroying the interference pattern, and the measurement results will reveal that the single particle indeed passes through both slits in a discontinuous way.

From the above analysis we can see, the first flaw in the demonstration of orthodox interpretation prevents people from assuming a new form of motion which is different from continuous motion in theory, while the second flaw further prevents people from detecting this kind of motion in experiment. Now we have remedied these two flaws, and accordingly the realistic motion picture of particles naturally appears. It is just the quantum motion of particles. As a result, the mystery of the double-slit experiment is finally unveiled.

In the double-slit experiment, the real process is that the single particle discontinuously passes through both slits. Concretely speaking, the particle is still in one of the two slits at each instant, but during a very small time interval the particle discontinuously moves throughout both slits and passes through them. Since the particle can pass through both slits in such a way, it will not only contain the information of one slit, but contain the information of both slits when arriving at the screen. Then it is intelligible that the double-slit interference pattern is not a simple mixture of two one-slit interference patterns. Certainly, in order to form

[17] See Section 4.3 for a detailed introduction of protective measurement.

the double-slit interference pattern, it not only requires that the particle passes through both slits, but also requires that the wave functions passing through the two slits superpose and interfere with each other. The latter is fulfilled by the law of quantum motion.

5.6.3 Schrödinger's Cat

Inspired by the paper of Einstein, Podolsky and Rosen (1935), Schrödinger (1935) expressed his dissatisfaction with quantum mechanics through a famous Gedanken experiment, which is later called Schrödinger's cat paradox. The experiment is described as follows.

"A cat is penned up in a steel chamber, along with the following diabolical device (which must be secured against direct interference by the cat); in a Geiger counter there is a tiny bit of radioactive substance, so small that perhaps in the course of one hour one of the atoms decays, but also, with equal probability, perhaps none; if it happens, the counter tube discharges and through a relay releases a hammer which shatter a small flask of hydrocyanic acid. If one has left this entire system to itself for an hour, one would say that the cat still lives if meanwhile no atom has decayed. The first atomic decay would have poisoned it. The ψ *-function of the entire system would express this by having in it the living and the dead cat (pardon the expression) mixed or smeared out in equal parts."*

Where is the paradox in Schrödinger's cat experiment? According to our macroscopic experience, the cat in the box can only be in a definite state, i.e., it is either living or dead, while according to quantum mechanics, the whole system in the box will be in a superposition of two states, in one of which the cat is living, in the other of which the cat is dead, and thus the cat will be in a bizarre superposition of living state and dead state. Consequently, it seems that quantum mechanics, which is generally regarded as a universal theory of Nature, actually contradicts our macroscopic experience. In this way, Schrödinger presented a paradox in his cat experiment. Either quantum mechanics is incapable of describing the macroscopic phenomena or our macroscopic experience is not real. In fact, what the paradox refers to is just the notorious measurement problem in quantum mechanics.

Most physicists assume a pragmatic point of view as to the dilemma in Schrödinger's cat experiment. They reckon that the principles of quantum mechanics may be replaced by those of classical mechanics at a certain level between the atoms and the measuring apparatus such as the Geiger counter, but they don't investigate where and how such transition happens. However, this problem must be solved sooner or later.

Now the theory of quantum motion provides a solution to the measurement problem. As a result, Schrödinger's cat paradox will be also solved. According to the law of quantum motion, a microscopic object such as an atom can be in a quantum superposition for a very long time. This forms the strange microscopic world. But as to a macroscopic object such as a cat, its quantum superposition has not been formed before the wave function collapse process finishes, thus the macroscopic object will be always in a definite state. This forms the familiar macroscopic world. In Schrödinger's cat experiment, the atom is indeed in the superposition of the initial state and the decay state. Since the energy distribution difference between these two states is very small, the collapse time of this superposition state is very long. But when this superposition is transferred through the quantum evolution to the macroscopic level such as the Geiger counter, the energy distribution difference between the states in the superposition will increase to a very large value. Then the collapse time of this superposition state will be very short, and thus the superposition state of the Geiger counter and the cat has not been formed before the collapse process finishes. As a result, the Geiger counter and the cat will always in a definite state in reality. The Geiger counter either registers a particle or registers no particles, and the cat is either living or dead.

Here we give a simple quantitative analysis of the above conclusion. A cat is a macroscopic complex open system, and the energy distribution difference ΔE between its living branch and its dead branch will be very large due to the environmental effects such as thermal energy fluctuations. According to the law of quantum motion, the collapse time $\tau_c \approx \frac{2\hbar E_P}{(\Delta E)^2}$ will be very short if ΔE is very large, and the quantum superposition of the living branch and the dead branch will collapse to one of the branches very soon. For example, for a cat comprising 10^{27} atoms, the energy distribution difference resulting from the thermal energy fluctuations will be $\Delta E \approx N^{1/2}kT \approx 900GeV$ for T = 300K, and the collapse time will be $\tau_c \approx 10^{-11}$ s. Then the quantum superposition of the cat has not been formed before the collapse process finishes, and thus the cat can only be in a living state or a dead state in reality. This solves the Schrödinger's cat paradox.

It can be seen that the characteristics of paradox are different for Schrödinger's cat experiment and double-slit experiment. Double-slit experiment mainly relates to the microscopic displays of quantum motion, and its paradox lies in the following fact, i.e., that classical mechanics cannot explain the microscopic phenomena, and thus it is not a complete theory. This fact has been widely accepted. Schrödinger's cat experiment mainly relates to the macroscopic displays of quantum motion, and its paradox lies in the following fact, i.e., that although quantum mechanics provides an accurate description of the microscopic world, it cannot account for the macroscopic phenomena. This indicates that quantum mechanics is not a complete theory either, and cannot provide a unified description for both the microscopic world and the macroscopic world yet. It is a pity that most people turn a deaf ear to this fact, and still wallow in the gratification brought by the great achievements of quantum mechanics today. However, the march of science can by no means stop, and the finding of quantum motion is just a small step on its road.

CHAPTER 6

Relativity in Discrete Space and Time

The theory of relativity is the other foundation stone of modern physics besides quantum theory. Although its formulation is very simple and graceful, it is not intelligible. The reason lies in that some postulates of the theory are not the results of a logical analysis, but only the direct representations of experience. For example, special relativity does not answer why the speed of light c is a universal constant. In this chapter we will try to provide a logical foundation for special relativity. It will be shown that the maximum and constancy of the speed of light may result from the discreteness of space and time. This explains one of the main postulates of special relativity. Moreover, we will propose a theory of relativity in discrete space and time, and discuss some of its inferences.

6.1 Why Is the Speed of Light Maximum and Constant?

The constancy of the speed of light is one of the main postulates of special relativity. According to this postulate, the speed of light in vacuum is the same in any inertial frame[18]. In addition, special relativity also requires that the speed of light is the maximum speed of objects in vacuum[19]. This may be the most mysterious and bewildering part of special

---

[18] Strictly speaking, the speed of light denotes the two-way average speed of light. In the framework of special relativity, the one-way speed of light is not a measurable physical quantity. For the convenience of discussions, we assume the usual Einstein simultaneity convention, i.e., stipulate the constancy of the one-way speed of light. It is actually an inference of the postulate of the isotropy of space.

[19] For a macroscopic object, its speed is defined as the speed of the apparent continuous motion of the object, which is the approximate display of quantum motion. For a microscopic object, its speed is defined as the group speed of the wave function describing its quantum motion.

relativity. Then why does there exist a maximum speed? And why is the maximum speed or the speed of light constant? Is there a deeper reason?

Since speed is essentially the ratio of space interval and time interval, it is a natural conjecture that the maximum and constancy of the speed of light may result from some undiscovered properties of space and time. In the following, we will argue that the discreteness of space and time may indeed result in the maximum and constancy of the speed of light.

Consider the continuous movement of an object in discrete space and time. If the object moves more than a space unit L_U such as $2L_U$ during a time unit T_U, then moving a space unit L_U will correspond to one half of the time unit T_U during the superluminal movement with speed 2c. Since the time unit T_U is the shortest measurable time interval in discrete space and time, and the duration of any change should be not shorter than the time unit T_U, such superluminal movement will be prohibited in discrete space and time[20]. Thus the object cannot move more than a space unit L_U during a time unit T_U, and its maximum speed will be the speed of light c, namely

$$v_{\max} = L_U / T_U = c \tag{6.1}$$

This result explains the maximum of the speed of light in terms of the discreteness of space and time. Consequently, the discreteness of space and time may be the physical cause of the maximum of the speed of light, at the same time, the maximum of the speed of light may have revealed that space and time is not continuous but discrete, in which the ratio of the space unit L_U and the time unit T_U is the speed of light c. By comparison, if space and time are continuous, then there exists no a characteristic time size and a characteristic space size, and

[20] An interesting inference is that there exist no tachyons in discrete space and time.

thus it seems very unnatural that there exists a characteristic speed c. Note that time interval and space interval are basic physical quantities, and speed is only a derivative physical quantity.

Now we will further demonstrate that the constancy of the speed of light also results from the discreteness of space and time. According to the principle of relativity, the time unit T_U and the space unit L_U should be the same in any inertial frame. If the minimum sizes of space and time are different in different inertial frames, then the inertial frames will be not equivalent, and there will exist a preferred Lorentz frame. This contradicts the principle of relativity. As a result, the speed c will be the maximum speed in any inertial frame. In order to demonstrate the constancy of the speed c, we need to analyze the movement of an object in two different inertial frames. Suppose an object moves with the maximum speed c in an inertial frame S. We need to find its speed in another inertial frame S'. Since the speed c is the maximum speed in any inertial frame, the speed of the object should be smaller than or equivalent to c in the inertial frame S'. If its speed is smaller than c, say $c/2$, in the inertial frame S', then due to the continuity of the velocity transformation, there must exist a speed larger than $c/2$ and a speed smaller than $c/2$ in the neighborhood of the speed $c/2$ in the inertial frame S', which correspond to the same speed smaller than c in the inertial frame S. This means that when the object moves with the above speed in the inertial frame S, it will have two possible speeds in the inertial frame S'. This is prohibited in logic. Thus we find that if an object moves with the maximum speed c in an inertial frame, it will also move with the same speed c in other inertial frames. This proves the constancy of the speed c in discrete space and time.

Since speed is a physical quantity derived from the ratio of space interval and time interval, it may be very natural that the characters of the speed of light can be further explained by the properties of space and time. As we have shown above, the maximum and constancy of the speed of light indeed result from the discreteness of space and time. This explains one of the main postulates of special relativity, and may provide a deeper logical foundation for special relativity.

6.2 Relativity in Discrete Space and Time

It is very odd that a fundamental theory like special relativity is based on a characteristic speed, namely the speed of light. Speed is evidently not a basic physical quantity, but derived from the ratio of space interval and time interval. This may have implied that the theory of relativity is just a makeshift, and will be replaced by a more fundamental theory based on the properties of space and time. The analysis in the last section reconfirms this conclusion. It shows that the characteristic speed appearing in special relativity can be further explained by the characteristic space-time sizes of discrete space and time. As a result, special relativity will be replaced by the theory of relativity in discrete space and time.

Relativity in discrete space and time has two postulates:

(1). the principle of relativity

(2). the constancy of the time unit T_U and the space unit L_U

The theory of relativity in discrete space and time is more fundamental than special relativity and general relativity, which are defined in continuous space and time. First, from its postulates we can derive the characteristic speed of light $c = L_U / T_U$, and deduce its constancy in any inertial frame. This will explain the most important postulate of special relativity, namely the constancy of the speed of light. Secondly, the discreteness of space and time permits the existence of a gravitational constant. We have the relation

$\kappa \equiv 8\pi G/c^4 = 2\pi L_U T_U/\hbar$, where κ is the Einstein gravitational constant, and G is the Newton gravitational constant. In continuous space and time where $T_U = 0$ and $L_U = 0$, we have $\kappa = 0$, and thus gravity does not exist. Thus it seems that gravity can only exist in discrete space and time, and general relativity should also be replaced by the theory of general relativity in discrete space and time.

In the following sections, we will discuss some inferences of the theory of relativity in discrete space and time.

6.3 On the Length Contraction

It can be seen that there exists an apparent contradiction between the constancy of the speed of light and the constancy of the space unit. It seems that the length contraction required by the constancy of the speed of light will permit no existence of a constant space unit. However, since the constancy of the speed of light results from the constancy of the time unit and the space unit, they should be consistent. The key lies in that the length contraction in special relativity will not hold true in discrete space and time. This means that the theory of relativity in discrete space and time will lead to a new length contraction formula. In the following, we will give a primary analysis of the length contraction in discrete space and time.

First, in order to satisfy the requirement of the constancy of the space unit, the length contraction factor must relate to the proper length, in other words, the length contraction factors should be different for different proper lengths. Then the length contraction factor can be written as follows:

$$\gamma_d^{-1} = \sqrt{1 - \frac{v^2}{c^2} + D(v, L_0)} \qquad (6.2)$$

where $L_0 \geq L_U$ is the proper length, $D(v, L_0)$ is a functional of v and L_0. Secondly, when $L_0 = L_U$, the constancy of the space unit requires that the observed length L in any inertial frame should be $L \equiv \gamma_d^{-1} L_0 = L_U$. Then we get:

$$\gamma_d^{-1} = \sqrt{1 - \frac{v^2}{c^2} + \frac{v^2}{c^2} D(L_0)} \tag{6.3}$$

where $D(L_U) = 1$. Here we omit the possibility that the length contraction factor contains the unnatural terms such as $(L_0 - L_U)^2$. Thirdly, when $v \to c$, the constancy of the space unit requires that the observed length L in the inertial frame moving with the velocity v should be $L \to L_U$ for any proper length L_0. Then we get:

$$\gamma_d^{-1} = \sqrt{1 - \frac{v^2}{c^2} + \frac{v^2}{c^2} (\frac{L_U}{L_0})^2} \tag{6.4}$$

This is the length contraction factor in discrete space and time. It is fully consistent with the requirement of the discreteness of space and time. For a microscopic particle, the length may be represented by its Compton wavelength. Then the length contraction factor can be rewritten as follows:

$$\gamma_d^{-1} = \sqrt{1 - \frac{v^2}{c^2} + \frac{v^2}{c^2} (\frac{E_0}{E_U})^2} \tag{6.5}$$

where $E_U = h / T_U$, E_0 is the energy of the particle.

It seems that if the space-time transformation still assumes the same linear form as Lorentz transformation, the new length contraction factor will be inconsistent with the constancy of the speed of light. One possibility is that the space-time transformation will be nonlinear in discrete space and time (cf. Amelino-Camelia 2002). However, even if the space-time transformation is still linear, such inconsistency may not exist when considering the quantum fluctuations of space-time in discrete space and time. The discreteness of space and time is essentially one kind of quantum property due to the universal existence of quantum motion. The space intervals and the time intervals will possess quantum fluctuations in discrete space and time. Moreover, the fluctuations will be larger in smaller space and time intervals. For example, such quantum fluctuations in discrete space and time can be revealed by the generalized uncertainty principle (GUP):

$$\Delta x \geq \frac{\hbar}{2\Delta p} + \frac{2L_P^{\ 2}\Delta p}{\hbar} = \frac{\hbar}{2\Delta p}[1 + \frac{L_P^{\ 2}}{(\hbar / 2\Delta p)^2}] \tag{6.6}$$

where $L_P = L_U / 2$. The first item in the right side represents a normal space interval in continuous space and time, and the second item in the right side represents the quantum fluctuation of the space interval resulting from the discreteness of space and time. As a result, the quantum fluctuation of a space interval $L = \frac{\hbar}{2\Delta p}$ is:

$$\Delta L = (\frac{L_P}{L})^2 L \tag{6.7}$$

Similarly, the quantum fluctuation of a time interval T is:

$$\Delta T = (\frac{T_P}{T})^2 T \tag{6.8}$$

where $T_P = T_U / 2$. In fact, a more general analysis of the measurement of space and time intervals shows that the quantum fluctuations of a space interval and a time interval may be larger (cf. Salecker and Wigner 1958). The results are:

$$\Delta L = (\frac{L_P}{L})^{2/3} L \tag{6.9}$$

$$\Delta T = (\frac{T_P}{T})^{2/3} T \tag{6.10}$$

Due to the existence of such quantum fluctuations of space intervals and time intervals, the constancy of the speed of light will not strictly hold true. Concretely speaking, the speed of light also possesses quantum fluctuation which relates to the traveling distance L of the light. Its average value is:

$$\Delta c \approx \frac{1}{2}(\frac{L_U}{L})^2 c \tag{6.11}$$

As a result, the new length contraction factor may be consistent with the constancy of the speed of light when considering the quantum fluctuations of space-time in discrete space and time.

6.4 An Inference of the Speed of Photon

In this section, we will discuss some implications of the length contraction factor in discrete space and time.

We assume the energy transformation $E = \gamma_d E_0$ still holds true in discrete space and time. This is a natural assumption. In fact, it is also consistent with the existing microscopic and macroscopic experience. When the speed of a microscopic particle is $v = c$, we have $\gamma_d^{-1} = E_0 / E_U$ according to the formula (6.5). Then the energy of the particle is $E = E_U \equiv E_P / 2$, where $E_P \approx 10^{19} Gev$ is the Planck energy[21]. Since the energy E_U is the maximum energy that a particle can have in discrete space and time, the particle moving with the maximum speed c will have the maximum energy, or vice versa. Note that the particle with any rest mass can reach the maximum speed c through acceleration. As an inference, the speed of the photon with the usual energy such as $1ev$ is not the maximum speed c, and is just

[21] Note that this result is independent of the rest mass of the particle. Especially, it is still valid when the rest mass is zero.

very close to the maximum speed c. Only the photon with the maximum energy can move with the maximum speed c. This result further implies that the rest mass of photon is not zero, but finite.

It can be seen that the above results are more natural than those of special relativity. In special relativity, the particle moving with the maximum speed c can have any energy such as a very small energy, and only the particle whose rest mass is zero can move with the maximum speed c, while the particle with a nonzero rest mass can never reach the maximum speed c. In fact, special relativity may have a deadly flaw here. According to the theory, in an inertial frame with speed c, the size of the whole world is zero, and its energy is infinite. Such a situation should not exist in physics. Thus special relativity will prohibit the existence of the inertial frames with speed c. On the other hand, photon moves with speed c in special relativity. Since photon is a real physical existence, the inertial frame comprising the photons with speed c should exist in physics. As a result, there exists an inherent inconsistency in special relativity. In addition, the existence of such photon frames will also require that photon should have a nonzero rest mass. In these frames, the photon is at rest and does exist, and thus it should have a nonzero energy.

In contrast with special relativity, the theory of relativity in discrete space and time has no the above inconsistency. It permits the existence of the inertial frames with speed c, and no infinity appears in physics when observing in these frames either. This analysis also supports the preceding conclusion that special relativity should be replaced by the theory of relativity in discrete space and time.

CHAPTER 7

A Theory of Quantum Gravity

Quantum theory and general relativity are the two main pillars of 20th-century physics. Quantum theory describes the quantum motion of objects without the influence of gravitation, and general relativity describes the gravitation between the objects undergoing classical motion. Whereas objects essentially undergo quantum motion and gravitation universally exists between the objects, a theory unifying quantum and gravity should be reasonably expected in order to provide a complete and consistent account of space-time and motion. Such to-be-found theory has been called quantum gravity or quantum general relativity. In this chapter we will suggest a possible way to unify quantum and gravity in terms of the quantum collapse in discrete space-time. We argue that it may provide a consistent theory of quantum gravity. In addition, we show that the mysterious dark energy may originate from the quantum fluctuations of the discrete space-time limited in our universe. This provides a possible support of our theory.

7.1 The Incompatibility between Quantum and Gravity

How to unify quantum and gravity turns out to be one of the hardest problems in physics. Quantum theory and general relativity are not only incomplete severally, but also incompatible with each other. Each of the two theories is unable to describe the quantum motion of objects under the influence of gravitation. Moreover, their views on how to describe such motion conflict with each other. Then why are quantum theory and general relativity incompatible?

As we know, quantum theory requires a presupposed definite space-time to define the quantum motion of objects, but according to general relativity, the space-time should be determined by the quantum motion of objects, and the resulting space-time is indefinite. Thus quantum theory and general relativity are incompatible when they are combined to describe the quantum motion of objects under the influence of gravitation. As a typical example, we consider a superposition state of different positions of a particle, say position A and position B. On the one hand, according to quantum theory, the valid definition of such a superposition requires the existence of a definite background space-time, in which position A and position B can be distinguished. On the other hand, according to general relativity, the space-time, including the distinguishability between position A and position B, cannot be predetermined, and must be dynamically determined by the superposition state. Since the different position states in the superposition determine different space-times, the space-time determined by the whole superposition state is indefinite. In such an indefinite space-time (i.e. the superposition of space-times), the quantum state and its evolution cannot be consistently defined. Concretely speaking, since position A and position B are in different space-times, and there exists no a pointwise identification between two space-times in general according to the principle of general covariance, position A and position B are undistinguishable in principle. As a result, the superposition state of position A and position B cannot be defined in physics. In addition, since the time-translation operator $\partial / \partial t$ is generally different in different space-times (e.g. the time-translation operators are related through the relation $\partial / \partial t' = \partial / \partial t + v \cdot \partial / \partial x$ in two space-times with a relative velocity v) (cf. Penrose 1996), the time-translation operator, which determines the evolution of the above superposition of two different space-times, cannot be consistently defined either.

In a word, quantum theory and general relativity are incompatible, and there exists a profound and fundamental conflict between the superposition principle in quantum theory and the principle of general covariance in general relativity (cf. Penrose 1996). Quantum theory rejects the superposition of different space-times, while the existence of such superposition seems to be an inevitable result of combining quantum theory and general relativity.

The incompatibility between quantum theory and general relativity indicates that there must exist a new physics beyond the description of the existing theories. Once we find and understand the new physics, the unification of quantum and gravity will be naturally achieved.

7.2 Quantum Collapse Helps to Reconcile Quantum and Gravity

As we think, the new physics may be the quantum collapse in discrete space-time. Quantum collapse will change the superposition of space-times to one of the definite space-times during a finite time interval. This provides a middle course to reconcile the conflicting quantum and gravity. The discreteness of space-time will ensure that quantum state and its evolution can still be consistently defined during the quantum collapse. This provides enough time for the reconciling process. Such quantum collapse in discrete space-time can accord with the existing experience.

We will explain the above reconciling method in more detail. When the space-times in the superposition have a difference much smaller than the minimum size of discrete space-time, they will be almost identical in physics[22]. This will ensure that quantum state and its evolution can be defined in such superposition of space-times. At the same time, the space-times in the superposition still have a bit of difference, and this will result in a very slow collapse of the superposition of space-times. For this kind of situation quantum theory is valid in a very precise way. When the space-times in the superposition have a difference near the minimum size of discrete space-time, they will be almost different in physics. This will result in a very quick collapse of the superposition of space-times. Such collapse will forbid the appearance of

[22] We note again that space intervals and time intervals will possess quantum fluctuations in discrete space-time, and thus the space-times with a difference smaller than the minimum size are not absolutely identical, but nearly identical in physics.

the superposition of space-times with a difference large than the minimum size of discrete space-time. For this kind of situation general relativity is valid in a very precise way.

In the middle situations, where the space-times in the superposition have a certain difference smaller than the minimum size of discrete space-time, quantum and gravity will continually interplay in a dynamical way. Quantum state will evolve in an approximately definite space-time in most time. When the quantum collapse process finishes, the superposition of space-times will collapse to one of the definite space-times, and then the quantum state will evolve in the new definite space-time. As time goes on, the quantum evolution will generate a new superposition of space-times, and quantum state will sequentially evolve in such an approximately definite space-time. Such processes will ceaselessly proceed due to the interplay between quantum motion and space-time. In short, space-time will be continuously changed by quantum evolution including quantum collapse, and quantum evolution will continuously proceed in such a dynamical space-time.

Once quantum theory and general relativity can be reconciled by the quantum collapse in discrete space-time, they will be naturally unified. It is really a big surprise that the hardest problems of quantum measurement and quantum gravity can be solved all together.

7.3 A Theory of Quantum Gravity

An immense amount of efforts have been made to unify quantum and gravity (cf. Rovelli, 2000). Yet although a great deal has been learned in the course of this endeavor, there is still no satisfactory theory. The present approaches still face severe problems, both technical and conceptual (cf. Butterfield and Isham 2001). It has been argued that it may be improper to quantize the gravitational field in a theory of quantum gravity (cf. Feynman 1995; Penrose 1996; Christian 2001; Weinstein 2001). The reasons include that the gravitational field is concerned with the structure of space-time that is fundamentally classical in nature etc. If it is indeed wrong to quantize the gravitational field, then how can we unify quantum and gravity in a consistent way? In this section, we will propose a consistent theory of quantum gravity in terms of the above reconciling method.

According to the analysis given in the last section, the theory unifying quantum and gravity should be a quantum field theory in the stochastic curved discrete space-time. For simplicity, we consider a real scalar quantum field ϕ of mass m propagating on the manifold $(M, g_{\mu\nu})$, which is a globally hyperbolic four-dimensional spacetime manifold M with metric $g_{\mu\nu}$. The theory can be formally formulated as follows:

$$G_{\mu\nu} = 8\pi G\langle \hat{T}^{R}_{\mu\nu}\rangle \tag{7.1}$$

$$\nabla_{\mu}\nabla^{\mu}\hat{\phi} - m^{2}\hat{\phi} - \hat{S}\hat{\phi} = 0 \tag{7.2}$$

The first equation is the semiclassical Einstein equation for the metric $g_{\mu\nu}$, where $G_{\mu\nu}$ is the Einstein tensor, G is the Newton gravitational constant, and $\langle \hat{T}^{R}_{\mu\nu}\rangle$ is the expectation value of the renormalized stress-energy operator in some physically acceptable state of the field on $(M, g_{\mu\nu})$. Note that both the stress tensor and the quantum state are functionals of the metric $g_{\mu\nu}$. The second equation is the evolution equation of the quantum field $\hat{\phi}$, where ∇_{μ} is the covariant derivative associated to the metric $g_{\mu\nu}$, and $\hat{S}$ is a stochastic evolution operator resulting in the dynamical collapse of the wave function or quantum field (cf. Chapter 3). Note that the field operator $\hat{\phi}$ is a functional of the metric $g_{\mu\nu}$ and the spacetime point x. A solution of the above equations consists of a spacetime $(M, g_{\mu\nu})$, a quantum field operator $\hat{\phi}[g]$ and a physically acceptable state $|\psi[g]\rangle$ for this field. It should be stressed that all quantities in the above equations of quantum gravity are defined in discrete space-time, and each of them contains the minimum space-time fuzziness in a direct or indirect way.

The stochastic evolution operator in the evolution equation of the quantum field will ensure the consistency of the above theory of quantum gravity. When the difference of the space-times corresponding to the branches of a quantum field is larger than the minimum size of discrete space-time, the superposition will instantaneously collapse to one of the branches with definite space-time due to the stochastic evolution. This ensures that there is no quantum superposition of physically different space-times, thus a quantum field and its evolution can always be consistently defined. As a result, general relativity and quantum field theory in curved spacetime (cf. Wald 1994; Ford 1997) can be derived from the above theory of quantum gravity as two approximate theories for respectively describing the motion of macroscopic objects and the motion of microscopic particles under the influence of gravitation.

In addition, when the difference of the space-times corresponding to the branches of a quantum field is smaller than the minimum size of discrete space-time, the space-times are almost identical in physics. This ensures that the semi-classical Einstein equation is an accurate equation for determining the background space-time of a quantum field. On the other hand, the space-times still have a bit of difference, and this will lead to the dynamical collapse of the quantum field. As a result, there also exist quantum fluctuations of space-time. The background space-time is continuously influenced by the quantum collapse, and undergoes intrinsic stochastic fluctuations. Note that the space-time undergoing such fluctuations is still definite at each moment.

In a word, quantum and gravity can be consistently unified with the help of the quantum collapse in discrete space-time. In this way, there is no quantized gravity in the usual meaning. Different from the semi-classical theory of quantum gravity, this theory may be a consistent fundamental theory of quantum gravity. Certainly, the properties of discrete space-time still need to be studied, though our analysis implies that space-time is not a dynamical entity possessing quantum properties.

7.4 How about Strings and Loops?

It is widely accepted that superstring theory and loop quantum gravity are the most promising alternatives to a complete theory of quantum gravity (cf. Polchinski 1998; Smolin 2001). However, if the above analysis is basically valid, then these two theories will be incomplete or even wrong as a theory of quantum gravity. In the following, we will give some brief comments on them.

First, superstring theory and loop quantum gravity are both incomplete due to the existence of the dynamical collapse of the wave function. They take it for granted that the linear quantum theory is absolutely right. However, as we have argued, the proper combination of quantum theory and general relativity will inevitably lead to the discreteness of space-time, and the discreteness of space-time may further lead to the dynamical collapse of the wave function. As a result, the linear quantum theory is not complete, and the final equation of quantum gravity must include a nonlinear stochastic evolution term describing the dynamical collapse process.

Secondly, owing to the existence of the dynamical collapse of the wave function, the quantum superposition of different space-times does not exist at all. This is also necessary for reconciling the fundamental conflict between the superposition principle in quantum theory and the principle of general covariance in general relativity. As a result, quantized gravity does not exist in the usual meaning. Since both superstring theory and loop quantum gravity insist that gravity should be quantized, and postulate the existence of the quantum superposition of different space-times, they may be simply wrong as a theory of quantum gravity. What they quantize may be not the real gravity or curved spacetime, but some kind of mathematical entity. Certainly, superstring theory and loop quantum gravity still deserve to be studied as useful mathematical models.

Feynman (1995) first noted that it is possible that gravity should not be quantized. He was also a strong opponent of superstring theory. Penrose (1996) further argued that the superposition of different space-times is physically improper. He is also an opponent of superstring theory and loop quantum gravity. History may finally confirm their forward-looking viewpoints.

7.5 A Conjecture on the Origin of Dark Energy

The recent observations show that our universe approximately contains 4% ordinary matter, 21% cold dark matter, and 75% dark energy (cf. Riess et al 1998, 2004; Perlmutter et al 1999; Spergel et al 2003; Allen et al 2004). In this section, we will propose a conjecture on the origin of the dark energy in our universe (cf. Gao 2005). The analysis indicates that the dark energy may originate from the quantum fluctuations of discrete space-time limited in our universe.

We assume our universe is a finite system limited by its event horizon in space. Whereas space-time is discrete when combining quantum theory and general relativity, the event horizon will contain finite area units, whose number is:

$$N = \frac{A}{{L_U}^2} = \frac{4\pi {L_H}^2}{4{L_P}^2} = \frac{\pi {L_H}^2}{{L_P}^2} \tag{7.3}$$

where A is the area of event horizon, $L_U \equiv 2L_P$ is the space unit in discrete space-time, L_P is the Planck length, and L_H is the event horizon of our universe. In addition, we assume the space-time limited in the event horizon undergoes quantum fluctuations according to the Heisenberg uncertainty principle in quantum theory, and its quantum fluctuation energy of one degree of freedom is:

$$\varepsilon \approx \frac{\hbar/2}{2L_H}c = \frac{\hbar c}{4L_H} \tag{7.4}$$

Since the quantum fluctuation of space-time of one degree of freedom corresponds to two area units in the two ends of the event horizon, the energy density of the quantum fluctuations of space-time in our universe is:

$$\rho_V \approx \frac{\varepsilon N/2}{4\pi L_H{}^3/3} = \frac{3\hbar c}{32L_p{}^2 L_H{}^2} = \frac{3c^4}{32GL_H{}^2} \tag{7.5}$$

By using the definition of event horizon $L_H = a(t)\int_t^\infty dt'/a(t')$, we can solve the Friedmann equation for our spatially flat universe. The evolution equation of Ω_V is:

$$\frac{d\Omega_V}{d\ln a} \approx \Omega_V(1-\Omega_V)(1+\frac{4}{\sqrt{\pi}}\sqrt{\Omega_V}) \tag{7.6}$$

where $\Omega_V \equiv \rho_V/\rho_c$, $\rho_c = 3H^2c^2/8\pi G$ is the critical energy density. The equation of state is:

$$w_V = -\frac{1}{3}\frac{d\ln\rho_V}{d\ln a} - 1 \approx -\frac{1}{3}(1+\frac{4}{\sqrt{\pi}}\sqrt{\Omega_V}) \tag{7.7}$$

By inputting the current value $\Omega_V \approx 0.75$, we can work out the equation of state:

$$w_V(z) \approx -0.98 + 0.24z + O(z^2) \tag{7.8}$$

In addition, we can also determine the current event horizon of our universe:

$$L_H \approx \frac{\sqrt{\pi}}{2\sqrt{\Omega_V}}H_0{}^{-1}c \approx 1.02H_0{}^{-1}c \tag{7.9}$$

This means that the current event horizon approximately satisfies the Schwarzschild relation $L_H = 2GM/c^2$, where $M = \rho_c 4\pi L_H{}^3/3$.

These theoretical results coincide with the recent cosmological observations (cf. Spergel et al 2003; Riess et al 2004; Allen et al 2004; Huterer and Cooray 2004), and a detailed analysis of the observational data further supports this dark energy model (Gong, Wang and Zhang 2004). For example, the recent results from Type Ia supernovae (SNe Ia) studies are $\Omega_X \approx 0.71 \pm^{0.03}_{0.05}$, $w = -1.02 \pm^{0.13}_{0.19}$, and $w_0 < -0.72$, $w' = 0.6 \pm 0.5$ with 95% confidence when constraining the analysis to $w_0 > -1$, where $w_0 \equiv w(z)|_{z=0}$ and $w' \equiv \frac{dw(z)}{dz}|_{z=0}$ (cf. Riess et al 2004). The analysis of Chandra measurements of the X-ray gas mass fraction shows $\Omega_X \approx 0.75 \pm 0.04$, $w_X = -1.26 \pm 0.24$, and $w_X < -0.7$ with 95% confidence when imposing the prior constraint $w_X > -1$ (cf. Allen et al 2004). In addition, the result of Huterer and Cooray (2004) mildly favors the dark energy models with w crossing -1. These results indicate that the above dark energy model is perfectly consistent with the observational data. As we think, the striking coincidence implies that the above quantum fluctuation energy of space-time is the only source of dark energy. As a direct consequence, the bare cosmological constant Λ_0 will be precisely zero, and the unknown scalar fields such as quintessence don't exist either (cf. Ratra and Peebles 1988; Scherrer 2004). This is a simple and natural assumption. If the above dark energy model is confirmed by further experiments, then using the relation $\Lambda = \frac{8\pi G}{c^2}\rho_V$, where Λ is the equivalent cosmological constant, we can obtain the Einstein equation for our universe:

$$R_{\mu\nu} - \frac{1}{2}Rg_{\mu\nu} + \frac{3\pi}{4}\frac{c^2}{L_H^{\ 2}(t)}g_{\mu\nu} = \frac{8\pi G}{c^4}T_{\mu\nu} \tag{7.10}$$

This equation will determine the fate of our universe.

We give some comments on the above analysis. It can be seen that the total quantum fluctuation energy of space-time is approximately $\frac{L_H}{L_P}E_P$, where E_P is the Planck energy.

By comparison, the total vacuum zero-point energy predicted by quantum field theory is approximately $\frac{L_H^3}{L_P^3}E_P$ when assuming the cut-off energy scale is the Planck energy. The former is nearly 120 orders of magnitude ($\frac{L_0^2}{L_P^2} \approx (\frac{10^{25}m}{10^{-35}m})^2 = 10^{120}$) smaller than the latter for the current universe. One half of the reduction comes from the fluctuation energy in each degree of freedom, and the other half comes from the whole number of degrees of freedom.

It is generally believed that the holographic form of dark energy is obtained by setting the UV and IR cutoff to saturate the holographic bound set by formation of a black hole (cf. Cohen, Kaplan and Nelson 1999; Horava and Minic 2000; Thomas 2002). Thus the dark energy can still come from the usual vacuum zero-point energy in quantum field theory. However, the above analysis implies that the usual vacuum zero-point energy may not exist. A simple calculation will show that even a holographic number of modes with the lowest frequency will give more vacuum zero-point energy than the observed dark energy. The lowest frequency of the vacuum zero-point energy limited in our universe is

$$E_1 = \frac{hc}{8L_H} \tag{7.11}$$

According to the holographic principle (cf. Bekenstein 1981; Hooft 1993; Susskind 1995), the whole number of degrees of freedom in our universe is

$$N_H = \frac{A}{4L_P^2} = \frac{\pi L_H^2}{L_P^2} \tag{7.12}$$

Then the vacuum zero-point energy density should satisfy the following inequality:

$$\rho_{VZE} \geq N_H E_1 = \frac{3\pi c^4}{16GL_H^2} \tag{7.13}$$

This requires $d \geq \sqrt{2}\pi/2$ in the holographic form of dark energy $\rho_V = \frac{3d^2c^4}{8\pi G L_H{}^2}$ (cf. Cohen, Kaplan and Nelson 1999; Horava and Minic 2000; Thomas 2002). Since the total energy in a region of the size L should not exceed the mass of a black hole of the same size, there should exist a theoretical upper bound $d \leq 1$. In addition, the result $d \geq \sqrt{2}\pi/2$ is also ruled out by the cosmological observations. This can be shown more directly from the equation of state:

$$w = -\frac{1}{3}(1 + \frac{2}{d}\sqrt{\Omega_V}) \geq -\frac{1}{3}(1 + \frac{2\sqrt{2}}{\pi}\sqrt{\Omega_V}) \tag{7.14}$$

By inputting the current fraction value $\Omega_V \approx 0.73$ we obtain $w_0 \geq -0.59$. This has been ruled out by the observational constraint $w_0 < -0.75$. Thus the dark energy may not come from the vacuum zero-point energy in quantum field theory, and the usual vacuum zero-point energy may not exist either (cf. Gao 2000; Rugh and Zinkernagel 2002). By comparison, the quantum fluctuations of space-time, whose energy density coincides with that of dark energy, may be the origin of dark energy. Since the quantum fluctuations of space-time may also be called quantum-gravitational vacuum fluctuations, the vacuum fluctuation energy still exists. It does not come from matter, but comes from space-time. In addition, since the density of vacuum fluctuation energy is inversely proportional to the square of the size of event horizon, it will be very large at the early stage of the universe. For example, it will be the Planck density $\rho_V \approx \frac{E_P}{L_P{}^3}$ in the Planckian era. If the evolution of the universe is dominated by such vacuum energy in the beginning, the inflation may be a natural result.

We note that our dark energy model will lead to dark energy behaving as phantom, and seems to violate the second law of thermodynamics during the evolution phase when the event horizon shrinks. However, the universe inside the event horizon is not an isolated system due to the universal existence of quantum processes such as Hawking radiation. Thus our dark energy model does not violate the second law of thermodynamics when considering the whole universe system. The universe inside the event horizon and that outside the event horizon will inevitably exchange energy and information due to the existence of quantum processes. This may also explain the non-conservation of dark energy inside the event horizon of our universe.

In comparison with the usual dark energy models, our dark energy model contains no adjustable parameters. It can be confirmed or disconfirmed more directly by experiments. According to the model, the density of dark energy is inversely proportional to the square of the event horizon of our universe. Its effects can be differentiated from those of the bare cosmological constant and the assumed scalar fields. If the model is valid, it may have some deep implications for discrete space-time and quantum gravity. In short, the quantum fluctuations of space-time do exist, and the dark energy has provided an experimental confirmation of discrete space-time. In addition, it also implies that the states of the quantum-gravitational vacuum may be essentially non-local. In other words, the local states are not independent each other, and can be decomposed into the independent non-local state bases. Undoubtedly, this dark energy model and the quantum fluctuations of space-time need to be further studied.

Lastly, we will predict a new quantum effect of black holes. If the quantum fluctuations of space-time limited in the event horizon of our universe do exist, then it should also exist between two black holes. This means that there will exist more quantum fluctuation energy or dark energy between two black holes. Consider two black holes with the same radius R. Let the distance between them be $L >> R$. The quantum fluctuation energy of space-time of one degree of freedom limited between them is $\varepsilon \approx \frac{\hbar c}{2L}$. The whole number of degrees of freedom of such fluctuations is $N \approx \frac{\pi R^2}{2L_p^{\ 2}}$. Then the whole quantum fluctuation energy of space-time between the two black holes is

$$E_{BHV} = N\varepsilon \approx \frac{\pi\hbar c R^2}{4L_p^{\ 2}L} = \frac{\pi R}{2L}E_{BH} \tag{7.15}$$

where E_{BH} is the energy of black hole. The energy density is

$$\rho_{BHV} \approx \frac{N\varepsilon}{\pi r^2 L} \approx \frac{\hbar c}{4L_p^{\ 2}L^2} \tag{7.16}$$

It is evident that the density of the quantum fluctuation energy between the black holes is much larger than that of the observed dark energy. Such energy can be detected in the local part of the universe such as the center of Milky Way. For example, the resulting repulsive acceleration of an object near one black hole is $a \approx \frac{2\pi c^2}{3L^2}r$, where r is the distance between the object and the black hole. The repulsive force equalizes the gravitational force of the black hole when $r \approx (RL^2)^{1/3}$. It is expected that this new quantum effect of black holes can be tested in the near future. Its finding will also provide an indirect confirmation of our dark energy model.

CHAPTER 8

Quantum Interactions

We have studied the motion of objects in space-time and the interplay of them. But in order to find the last law of motion, we must further study the interactions between objects. Then how do the objects interact with each other? A theory named quantum field theory has been founded to describe the interactions between objects. Although the predictions of the theory accurately coincide with experiment, it appears that the theory has not been well understood. What on earth is a quantum field? Is it a real physical field? Or is it only a description of the relativistic motion of many particles? In this chapter we will analyze the quantum mechanism of interactions, and give the real physical picture of quantum field.

8.1 Why Do the Particles Create and Annihilate?

The existence of a particle is eternal in quantum mechanics. In quantum field theory, however, a particle can create and annihilate, or we can say, a particle has its life. The creation and annihilation processes of particles are the most amazing processes in quantum field theory. Such processes have also been found in experiment. Then why do the particles create and annihilate? And how to understand the creation and annihilation processes of particles?

As we know, the interaction between two particles is via a classical field in (nonrelativistic) quantum mechanics. The results of the interaction are only the motion changes of the particles and the corresponding change of the classical field. As a result, the existence of a particle is eternal. However, in (relativistic) quantum field theory, since there exist no classical fields, and the basic existent forms of matter are only the particles satisfying relativistic motion equations, the local interaction between two particles can only be implemented by the transfer of other particles, while such transfer necessitates the creation and annihilation of the transfer particles. Consequently, in order to generate the interaction between particles, the particles must create and annihilate.

Then why must interactions exist? In reality, all properties of a particle are defined relative to other particles, in other words, they are defined through the interaction between the particle with other particles. For example, the mass of a particle is defined as the ratio of the external force experienced by the particle (i.e. the interaction between the particle and other particles) and its acceleration. If there exist no interactions, then a particle will devoid of all properties, and will no longer exist either. Accordingly interactions are the precondition of the existence of particles. Particles must interact in order to exist, and the interactions must be implemented through the creation and annihilation of particles. As a result, particles must create and annihilate in order to exist. This reconfirms an old credo, i.e., that being exists at a cost of finite life. For each particle, the beginning of life is its creation, and the end of life is its annihilation. This fact is clearly described by the interaction terms in quantum field theory. For example, the interaction term $-e\overline{\psi}\gamma^{u}\psi A_{u}$ describes the local interaction between an electron and a photon, and denotes that an electron and a photon annihilate and then a new electron creates.

Particles must create and annihilate. Then where do the particles create from? And where do the particles annihilate to? The place that the particles create and annihilate in is the omnipresent vacuum. In quantum field theory, the vacuum state naturally exists as the state in which no particles exist, and it is also the lowest energy state in the physical world. As a state with the lowest energy, the vacuum provides the final place for the annihilated particles. As a result, the number of particles is no longer conserved in quantum field theory.

Lastly, we note that a basic interaction charge can be constructed by using the space unit L_U, the time unit T_U and the unit of motion $\hbar$. The expression can be written as $C_I = \sqrt{E_U \cdot L_U} = \sqrt{\hbar L_U / T_U} = \sqrt{\hbar c}$, and the interaction potential energy is in inverse proportion to the distance between the charges. The charge of the electromagnetic interaction can be expressed as a multiple of the basic charge, namely $e = \sqrt{\alpha \cdot \hbar c} = \sqrt{\alpha} \cdot C_I$, where $\alpha \approx 1/137$ is the fine structure constant. Similarly, the charges of weak interaction and strong interaction can both be expressed as a certain multiples of the basic charge. The multiple denotes the coupling strength of the interaction, and may be determined by the interplay of the charge and the actual vacuum.

8.2 What Is Quantum Field?

Even though quantum field theory has gained great success in its accurate consistency with experiment, it seems that the theory has not been well understood yet. In this section, we will try to understand the quantum field theory[23], and give the real physical picture of quantum field. Our analysis will show that quantum field is not a kind of physical field, and what quantum field theory describes is the motion of many particles in the relativistic domain, which includes the creation and annihilation of particles as special motion processes.

The core of quantum field theory is its operator field formalism. As we know, there exist only two kinds of operators in operator field. One is creation operator, which is denoted by a_k^+ in momentum space, and the other is annihilation operator, which is denoted by a_k in momentum space. On the one hand, the introduction of creation and annihilation operators has its mathematical validity. They come from the Fock space description of many-particle state, which is equivalent to the real space description of many-particle state in mathematics. On the other hand, the introduction of creation and annihilation operators also has its physical validity. There indeed exist the creation and annihilation processes of particles, and we can more directly describe these two processes through the creation and annihilation operators.

---

[23] For a detailed analysis of the conceptual developments of quantum field theory, see Cao (1997).

According to the definitions of quantum field theory, quantum field is the creation and annihilation operators in real space. It can be generally written as follows:

$$\psi^{+}(x,t)=\psi_a^{+}(x,t)+\psi_b(x,t) \tag{8.1}$$

and

$$\psi(x,t)=\psi_a(x,t)+\psi_b^{+}(x,t) \tag{8.2}$$

where $\psi_a^{+}(x,t)$ and $\psi_a(x,t)$ denote the creation and annihilation operators of a particle in the space-time point (x,t), $\psi_b^{+}(x,t)$ and $\psi_b(x,t)$ denote the creation and annihilation operators of its antiparticle in the space-time point (x,t). Since these operators relate to space and time, they are visually called operator field or quantum field. The physical meaning of these operators lies in their actions on the many-particle states. For example, when $\psi^{+}(x_1,t_1)$ acts on the vacuum state $|0>$, we have:

$$\psi^{+}(x_1,t_1)|0>=\psi_a^{+}(x_1,t_1)|0>+\psi_b(x_1,t_1)|0>=\delta_a(x-x_1) \tag{8.3}$$

This means that the action of $\psi^{+}(x_1,t_1)$ on the vacuum state is to create a particle in position x_1 at instant t_1.

The creation and annihilation operators in real space can be expanded in momentum space, and there exists a one-to-one relation between them. This relation is determined by the properties of the particle such as charge and spin etc. For the charged particle with spin 0, this relation can be written as follows:

$$\psi(x,t)=\int d\tilde{k}\,[a(k,t)e^{-i(kx-\omega t)}+b^{+}(k,t)e^{i(kx-\omega t)}] \tag{8.4}$$

where $a(k,t)$ denotes the annihilation operator of the particle in momentum space, $b^{+}(k,t)$ denotes the creation operator of its antiparticle in momentum space. From the expansion form of quantum field in momentum space, we can further see that there exist two kinds of physical processes in the description of quantum field. One is the free relativistic motion of the particles (including the particle and its antiparticle). It is described by the momentum eigenstates of the particles such as $e^{-i(kx-\omega t)}$. The forms of the momentum eigenstates are determined by the properties of the particles. The other is the creation and annihilation processes of the particles (including the particle and its antiparticle), which can be regarded as the relativistic motion of the particles under the influence of interaction. These processes are described by the creation and annihilation operators of the particles in momentum space such as $a(k,t)$. Moreover, the physical definitions of the creation and annihilation operators will determine their nontrivial commutation and anticommutation relations in quantum field theory (see the following Appendix), which provide the so-called basis of field quantization. Accordingly, the motion of particles is still the true physical reality hiding behind quantum field, and what quantum field theory describes is the motion of many particles in the relativistic domain.

We can summarize the relations between quantum field theory and the related theories as follows.

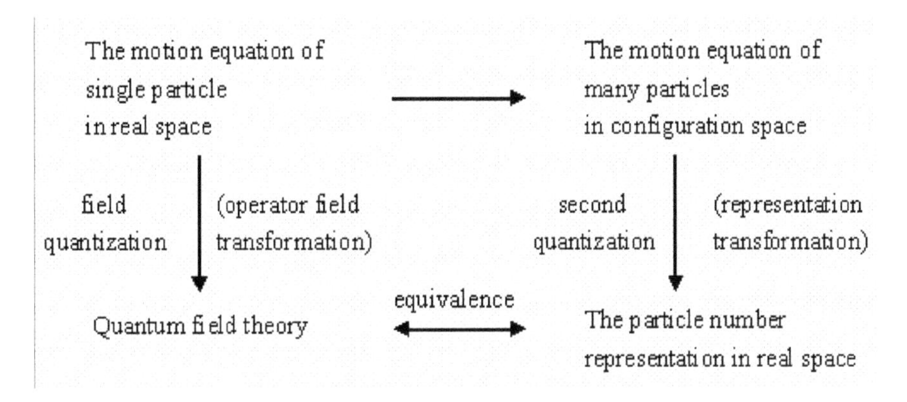

Figure 7 The relations between quantum field theory and the related theories

From the previous figure, we can see that quantum field theory is equivalent to the particle number representation in three-dimensional space, which can be obtained from the many-particle motion equation in 3N-diemnsional configuration space through the representation transformation. Thus what quantum field theory describes is essentially the motion of many particles. This conclusion is valid in both nonrelativistic and relativistic domains. Note that Teller (1995) had argued that the Fock space representation gives the real physical interpretation of the formalism of quantum field theory.

Now we can finally answer the question 'what is quantum field?' Quantum field is not a physical field, but a mathematical description of the relativistic motion of many particles. The physical reality hiding behind quantum field is still the quantum motion of particles, which includes the creation and annihilation of particles as special motion processes.

Appendix: An analysis of the creation and annihilation operators

In order to understand the quantum field theory, we need to analyze the creation and annihilation operators a_k^+ and a_k in more detail. Here we will demonstrate that their commutation and anticommutation relations assumed by the theory are actually determined by their physical definitions. According to the definitions of creation and annihilation operators, we have $a_k^+|n_k\rangle = g(n_k)|n_k+1\rangle$, $a_k|n_k\rangle = f(n_k)|n_k-1\rangle$, and when $n_k=0$, $a_k|n_k\rangle=0$. We first assume a_k^+ and a_k are commutative, namely $[a_k^+, a_k] = 0$. This leads to the following relation for any n_k:

$$g(n_k)\, f(n_k+1) = f(n_k)\, g(n_k-1) \tag{8.5}$$

By using the definition $a_k|0\rangle=0$ we find $f(0)=0$. Then the above relation requires $g(0)\, f(1)=0$. From the definition $a_k^+|0\rangle = g(0)|1\rangle$ we know $g(0) \neq 0$, then we get

$f(1)$=0. By using the above relation again and again, we can further get $f(n_k)$=0 for any n_k, and thus we have a_k=0. This result evidently contradicts the existence of the annihilation operator a_k. Thus the assumption that a_k^+ and a_k are commutative is wrong. Then we prove that the creation and annihilation operators are not commutative, namely $[a_k^+, a_k] \neq 0$. When assuming the commutative relation is irrelevant to the state $|n_k>$, we can further get the commutative relation $[a_k^+, a_k]=1$. In a similar way, we can also prove that the creation and annihilation operators are not anticommutative, namely $\{a_k^+, a_k\} \neq 0$. When assuming the anticommutative relation is irrelevant to the state $|n_k>$, and $|n_k>$ can only assume |0> and |1>, we can get the anticommutative relation $\{a_k^+, a_k\}=1$. In addition, since the creation and annihilation operators of different states are irrelevant, they are evidently commutative or anticommutative, i.e., when $k' \neq$ k, we have $[a_k^+, a_{k'}]=0$ or $\{a_k^+, a_{k'}\}=0$. Similarly, the creation operators are commutative or anticommutative, and the annihilation operators are commutative or anticommutative, i.e., $[a_k^+, a_{k'}^+]=0$ and $[a_k, a_{k'}]=0$ or $\{a_k^+, a_{k'}^+\}=0$ and $\{a_k, a_{k'}\}=0$.

The above analysis clearly shows that the nontrivial commutation and anticommutation relations between the creation and annihilation operators are determined by their physical definitions, especially by the intrinsic opposition characters of the creation and annihilation processes. Thus the commutation relation between the creation and annihilation operator is different from that between position and momentum in quantum mechanics. As a result, the so-called field quantization is not a kind of quantization in the usual meaning.

Lastly, we note that the above commutation and anticommutation relations between the creation and annihilation operators are the simplest relations. It is possible that the commutative and anticommutation relations are relevant to the state, and thus there exist more complex relations between the creation and annihilation operators. These relations may be necessary for the unification of fermions and bosons.

CHAPTER 9

Quantum Non-locality

In this chapter we will analyze the perplexing quantum non-locality in terms of the realistic picture of quantum motion. We will demonstrate that the collapse process of the wave function requires the existence of a preferred Lorentz frame. A possible method to detect the preferred Lorentz frame is also proposed. This provides a solution to the problem of the incompatibility between quantum non-locality and special relativity.

9.1 An Analysis of Quantum Non-locality in Terms of Quantum Motion

Quantum motion is essentially discontinuous in real space, thus it naturally contains non-local processes. As we know, the physical picture of the quantum motion of a single particle is as follows. A particle stays in a space unit L_U during a time unit T_U. Then it will still stay there or stochastically appear in another space unit L_U, which may be very far from the original region, during the next time unit T_U. During a time interval much larger than the time unit T_U, the particle will discontinuously move throughout the whole space with a certain position measure density $\rho(x,t) = |\psi(x,t)|^2$. Since the distance between the space units occupied by the particle during the neighboring time units may be very large, the motion process is evidently non-local. It seems that space does not exist for the particle undergoing quantum motion.

In the following, we will mainly analyze the non-local characters of the entangled state of particles in terms of the realistic picture of quantum motion. This analysis will make people see the weird displays of quantum non-locality more clearly. Consider a TPES (Two-Particle

Entangled State) $\psi_1\varphi_1 + \psi_2\varphi_2$, where the states $\psi_1\varphi_1$ and $\psi_2\varphi_2$ are separated very far in real space. The physical picture of this TPES is as follows. Particles 1 and 2 are in state $\psi_1\varphi_1$ during a time unit T_U, and then they will still stay in this state or be in state $\psi_2\varphi_2$ during the next time unit T_U. During a very short time interval which is still much longer than the time unit T_U, the two particles will discontinuously move throughout the states $\psi_1\varphi_1$ and $\psi_2\varphi_2$ with the same measure density 1/2. Note that particles 1 and 2 being in state $\psi_1\varphi_1$ means that particle 1 is in state ψ_1 and particle 2 is in state φ_1. Similarly, particles 1 and 2 being in state $\psi_2\varphi_2$ means that particle 1 is in state ψ_2 and particle 2 is in state φ_2. It can be seen that the motion process of the quantum whole comprising particles 1 and 2 is evidently non-local due to the essential discontinuity of quantum motion.

Now we make a measurement of particle 1 in a local region. Let the initial state of the measuring apparatus be χ_0. When the measuring apparatus interacts with particle 1 in a local region, the state of the measuring apparatus will be entangled with the TPES, and the state of the whole system will turn to be $\chi_1\psi_1\varphi_1 + \chi_2\psi_2\varphi_2$. Since the measuring apparatus introduces a very large energy distribution difference between the two branches $\chi_1\psi_1\varphi_1$ and $\chi_2\psi_2\varphi_2$, the state of the whole system will soon collapse to one of the branches $\chi_1\psi_1\varphi_1$ or $\chi_2\psi_2\varphi_2$ according to the law of quantum motion. As a result, the quantum whole is disassembled into three independent parts: the measuring apparatus, particle 1 and particle 2. At the same time, the initial two-particle quantum whole no longer exists either, and its state collapses to $\psi_1\varphi_1$ or $\psi_2\varphi_2$, i.e., the state of particle 1 collapses to ψ_1 or ψ_2, and the state of particle 2 collapses to φ_1 or φ_2. It can be seen that the local measurement of particle 1 brings a non-local influence on the two-particle quantum whole, and thus brings a non-local influence on particle 2. The distance between the measuring apparatus and particle 2 can be arbitrarily large, and the collapse time can be arbitrarily short in principle. Accordingly the apparent speed of such an influence can be close to infinite. Note that there exists no causal influence between the state changes of particle 1 and particle 2, and they both result from the local measurement, which is the common cause.

Since the measurement simultaneously results in the collapse of the states of particle 1 and particle 2, there also exists a non-local correlation between the collapse states of particle 1 and particle 2. Concretely speaking, if the state of particle 1 collapses to ψ_1, then the state of particle 2 must collapse to φ_1, whereas if the state of particle 1 collapses to ψ_2, then the state of particle 2 must collapse to φ_2. Since the state changes of particle 1 and particle 2 are simultaneous, and the measurements of their collapse states can be made in space-like separated regions, the correlation between the measurement results of particle 1 and particle 2 may be still non-local. Such non-local correlation can be revealed through some inequalities such as Bell's inequality (Bell 1964), and has been confirmed by experiment (cf. Aspect 1982). However, as Bell (1987) said, "The scientific attitude is that correlations cry out for explanation." The above analysis may give a true physical explanation of quantum non-local correlations in terms of quantum motion.

Lastly, we note that there still exist classical non-local correlations between the classical events happening in space-like separated regions. Different from quantum non-local correlations, the correlations are completely pre-determined by some common causes in the past via the classical influences with finite speed, and there exist no randomness and non-local influences in the measurement of such correlations.

9.2 The Existence of Preferred Lorentz Frame

We have analyzed quantum non-locality in a given Lorentz frame. In this section, we will analyze its displays in different Lorentz frames. It will be shown that, in order to avoid causal loops, there must exist a preferred Lorentz frame for defining the actual causal relations of the non-locally correlating events. As a result, quantum non-locality does not satisfy Lorentz invariance in essence.

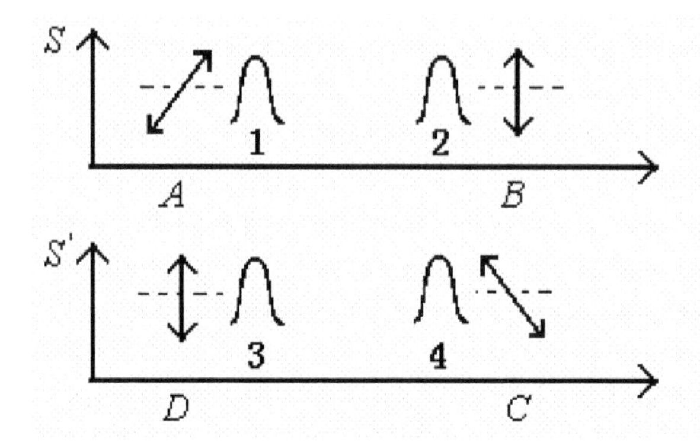

Figure 8 Quantum non-local experiments in two Lorentz frames

Consider two Lorentz frames S and S' in relative motion. There are two independent TPESs in the two frames respectively. In frame S, particle 1 is measured in position x_A at instant t_A, and particle 2 is measured in position x_B at instant t_B. Measurement A is earlier than measurement B, namely $t_A < t_B$, and they satisfy the space-like separation condition $|x_B - x_A| > c|t_B - t_A|$. In frame S', particle 3 is measured in position x_D' at instant t_D', and particle 4 is measured in position x_C' at instant t_C'. Measurement C is earlier than measurement D, namely $t_C' < t_D'$, and they satisfy the space-like separation condition $|x_D' - x_C'| > c|t_D' - t_C'|$. In addition, there exists a connection between the frames S and S', i.e., the result of measurement B determines the setting of measurement C

(e.g. the setting of angle for spin measurement) in a classical way, and the result of measurement D determines the setting of measurement A in a classical way. This leads to the time order relations $t_B < t_C$ and $t_D < t_A$ in frame S. Then the time order of the four measurements is $t_D < t_A < t_B < t_C$ in frame S, and the time order of the measurements C and D is $t'_C < t'_D$ in frame S'. Such time order of events can exist as a result of the Lorentz space-time transformation in special relativity. As we know, there exist no causal connections between the events A and B and between the events C and D in classical domain, thus the mixed time order loop of these four events (namely $t_D < t_A < t_B < t_C$ and $t'_C < t'_D$) in the frames S and S' does not result in a causal loop.

Now we consider the quantum non-local influences in the above experiment. We assume the collapses of the wave functions of the particles in TPES are simultaneous in all Lorentz frames, i.e., the simultaneity of the wavefunction collapse possesses Lorentz invariance. In frame S, since measurement A is earlier than measurement B, and the collapse process of the TPES is simultaneous in positions x_A and x_B, the setting selection of measurement A will non-locally influence the result of measurement B, i.e., there exists a non-local causal connection $A \rightarrow B$. In the similar way, there also exists a non-local causal connection $C \rightarrow D$. Then there exists a causal connection between the events in every segment of the above mixed time order loop. When we properly select the classical connections, say $B \rightarrow C$ and $D \rightarrow \overline{A}$, the mixed time order loop will form a causal loop, say $A \rightarrow B \rightarrow C \rightarrow D \rightarrow \overline{A}$. In the framework of present quantum theory, the quantum non-local connections $A \rightarrow B$ and $C \rightarrow D$ are both probabilistic. Thus the above causal loop will appear with a certain probability[24]. This is also forbidden in logic. Accordingly the preceding assumption is wrong, i.e., the simultaneity of the wavefunction collapse does not possess Lorentz invariance[25]. This will lead to the existence of a preferred Lorentz frame. The preferred Lorentz frame can be defined as the inertial frame in which the collapse of the wave function happens simultaneously in the positions occupied by the wave function. In other Lorentz frames, the collapse process is not simultaneous, and its time order is determined by the time order in the preferred Lorentz frame through the Lorentz space-time transformation.

Once there exists a preferred Lorentz frame, the mixed time order loop will not form a causal loop in the above experiment. The reason is that one of the quantum non-local connections $A \to B$ or $C \to D$ will not exist. Concretely speaking, its causal connection direction will be reversed, say $A \to B$ turns to $B \to A$. In the preferred Lorentz frame, all quantum non-local influences are simultaneous, and the causal relation between the non-locally correlating events is actually and exclusively determined. But in other Lorentz frames, the quantum non-local influences will be no longer simultaneous, and the causal relation between the non-locally correlating events will no longer accord with their time orders, but be only determined by their time orders in the preferred Lorentz frame. This guarantees that causes always come before effects. Thus it is evident that there will no longer exist causal loops for the quantum non-local influences.

In a word, we have demonstrated that the quantum non-local influence resulting from the collapse of the wave function does not satisfy Lorentz invariance, and thus leads to the existence of a preferred Lorentz frame. It can be seen that this conclusion is a special case of a

---

[24] See Percival (1998) for a detailed analysis.

[25] Strictly speaking, what we have demonstrated is that the preceding assumption is incompatible with special relativity, i.e., the simultaneity of the wavefunction collapse and the one-way speed of light cannot both possess Lorentz invariance. It is actually a demonstration of the incompatibility between quantum non-locality and special relativity. Here we implicitly assume the constancy of the one-way speed of light.

general conclusion, which says that only one speed value is permitted to be constant in any Lorentz frame, and if there are more, then causal loops can be produced. In the above experiment, if we assume that two speed values, the speed of light (i.e. one-way speed of light) and infinite speed, are both constant in any Lorentz frame, then causal loops will naturally appear. Our solution is to assume the validity of the constancy of one-way speed of light. Then the constancy of infinite speed should be rejected. Thus quantum non-local influence, whose speed is infinite, will naturally result in the existence of a preferred Lorentz frame.

It is worth noting that the preferred Lorentz frame does not require the existence of some kind of background field or quantum ether. The existence of preferred Lorentz frame completely results from the non-locality of the wavefunction collapse, which is one of the essential characters of quantum motion. In addition, whereas the time order of the wavefunction collapse is isotropic in the preferred Lorentz frame, it can be reasonably guessed that the preferred Lorentz frame may be the cosmos frame, in which the microwave background radiation is isotropic, and the temperature of radiation provides an absolute measure of the cosmos time.

Lastly, we stress that the absolute validity of the principle of relativity must be limited due to the existence of preferred Lorentz frame. The principle of relativity does not hold true for the collapse of the wave function. It should be mentioned that Einstein, the founder of special relativity, also realized the possible limitation of the principle of relativity. He said, "As long as one was convinced that all natural phenomena were capable of representation with the help of classical mechanics, there was no need to doubt the validity of this principle of relativity. But in view of the more recent development of electrodynamics and optics, it became more and more evident that classical mechanics affords an insufficient foundation for the physical description of all natural phenomena. At this juncture the question of the validity of the principle of relativity became ripe for discussion, and it did not appear impossible that the answer to this question might be in the negative" (Einstein 1920).

9.3 How to Detect the Preferred Lorentz Frame?

The incompatibility between quantum non-locality and special relativity has been demonstrated from different points of view (cf. Bell 1986; Hardy 1992; Squires 1992; Percival 1998, 2000; Suarez 2000a, 2000b; Scarani et al 2000; Gao 2000). It has been argued that the collapse of the wave function can result in the appearance of quantum non-locality, and further requires the existence of a preferred Lorentz frame. The analysis given in the last section reconfirms this conclusion. Then how to detect the preferred Lorentz frame will be a pressing problem. In this section, we will propose a possible method to detect the preferred Lorentz frame in terms of the theory of quantum motion. According to the theory, the collapse time of the wave function relates to the velocity of the experimental frame relative to the preferred Lorentz frame. Thus the preferred Lorentz frame can be detected through measuring the collapse time of the wave function.

According to the theory of quantum motion, the collapse of the wave function is described as a dynamical process, and the nonrelativistic collapse time formula for a two-level system is:

$$\tau_c = \hbar E_p /(\Delta E)^2 \tag{9.1}$$

where $\hbar$ is the Planck constant divided by 2π, E_p is the Planck energy, ΔE is the squired energy uncertainty of the state. Here we omit a finite dimensionless coefficient which may be involved in the collapse time formula. Assume this nonrelativistic collapse time formula is valid in a frame *S* in the relativistic domain. Then using the Lorentz transformations, we can get the relativistic collapse time formula in frame *S′* with velocity v relative to frame *S* is:

$$\tau_c = (1 - uv/c^2)^{-3}(1 - v^2/c^2)^{3/2}[\hbar E_p /(\Delta E)^2] \tag{9.2}$$

where u is the group velocity of the measured wave function in frame *S*. This formula contains the terms relating to the velocity v of the experimental frame. This evidently violates the principle of relativity. We may define the frame in which the collapse time formula (9.1) is valid as the preferred Lorentz frame. Thus according to the theory of quantum motion, the collapse law provides a method to detect the preferred Lorentz frame.

In general, we can measure the collapse time of the wave function through measuring the change of the interference between the corresponding collapse branches in the initial wave function. The change of the interference effect due to collapse can be generally formulated as $\rho(t) = \rho(0)e^{-t/\tau_c}$. The main technology difficulty of this general approach is to exclude the influence of environmental decoherence. Marshall et al (2003) proposed an experimental scheme for testing the existence of the dynamical collapse of the wave function. Their experiment may also be used to detect the preferred Lorentz frame.

In the following, we will further propose a possible experimental scheme to probe the preferred Lorentz frame, which can overcome the above difficulty. Consider the K_L^0 meson decay process. The state of K_L^0 meson can be written as follows:

$$|K_L^0> = \frac{1}{\sqrt{2}}(|K^0> - |\overline{K}^0>) = \frac{1}{\sqrt{2}}(|s>|\overline{d}> - |d>|\overline{s}>) \tag{9.3}$$

According to the theory of quantum motion, in the above collapse time formula we have $\Delta E \approx 100Mev$, which is approximately the mass difference of s quark and d quark[26], and the corresponding collapse time τ_c is nearly $10^{-3}s$ in the preferred Lorentz frame. Since the K_L^0 meson decay process is very complex, it is very difficult to directly measure the collapse time through measuring K^0 and $\overline{K}^0$. Fortunately, we may indirectly find the collapse time by measuring the CP branching ratio of K_L^0 due to the involved CP violation.

---

[26] Here we assume the wave functions of s quark and d quark in the K_L^0 meson are approximately separated in space.

The initial state $|K_L^0>$ is a CP eigenstate, while the collapse states $|K^0>$ and $|\overline{K}^0>$ are not CP eigenstates. This indicates that the collapse of the wave function is the cause of the CP violation in the K_L^0 meson decay process. Thus we can work out the CP branching ratio of K_L^0 meson in terms of the collapse time τ_c and the mean life τ of K_L^0 meson[27]:

$$\gamma = \frac{\gamma_c \tau}{\tau_c} \tag{9.4}$$

where γ is the CP branching ratio of K_L^0 meson, γ_c is the collapse branching ratio which equals to $1/2$[28]. Then we can probe the preferred Lorentz frame through measuring the CP branching ratio of K_L^0 meson.

According to the relativistic collapse time formula (9.2), when the group velocity of the measured wave function is close to the speed of light, and the velocity of the experimental frame relative to the preferred Lorentz frame is much smaller than the speed of light, the collapse time formula in the first rank of v is:

$$\tau_c \approx (1+3v/c)[\hbar E_p/(\Delta E)^2] \tag{9.5}$$

Then the difference of the collapse times measured in different experiment frames is:

$$\Delta\tau_c \approx (3\Delta v/c)\tau_c \tag{9.6}$$

---

[27] Here we assume the collapse of the wave function is the only cause of the CP violation in the K_L^0 meson decay process.

[28] This relation was first noticed by Fivel (1997a, 1997b). It is consistent with the present experiment data of both K_L^0 and K_S^0. Using this relation we can get the CP branching ratio of K_S^0 $\gamma_S \approx 0.52\times10^{-5}$. This value accords with the present measurement result $\gamma_S < 1.4\times10^{-5}$ (cf. Eidelman et al 2004).

Using the relation (9.4) we can further get the corresponding difference of the CP branching ratios of K_L^0 meson:

$$\Delta\gamma \approx -(3\Delta v / c)\gamma \tag{9.7}$$

Assume the preferred Lorentz frame is the cosmos frame (CMB-frame), in which the cosmic background radiation is isotropic. Considering the speed of the Earth relative to the CMB-frame $v \approx 370 km/s$ and the maximum difference of the revolution speed of the Earth relative to the Sun $\Delta v \approx 60 km/s$, the maximum difference of the CP branching ratios of K_L^0 meson measured in different times (e.g. spring and fall respectively) on the Earth will be

$$|\Delta\gamma| \approx |3\Delta v/c|\gamma \approx 6\times10^{-4}\gamma \tag{9.8}$$

The recent measured CP branching ratio of K_L^0 meson is $(2.858\pm0.023)\times10^{-3}$ (cf. Eidelman et al 2004). Thus if the measurement could be made more accurately, then the preferred Lorentz frame can be detected through measuring the seasonal change of the CP branching ratio of K_L^0 meson on the Earth. This conclusion also holds true for the other decay processes involving CP violation such as the K_S^0 meson decay. It is expected that such measurement can be accomplished in the near future (cf. Kleinknecht 2003).

The above method for detecting the preferred Lorentz frame also provides a proof of the existence of preferred Lorentz frame. Moreover, the analysis reconfirms the conclusion that the preferred Lorentz frame is determined by the collapse law of quantum motion. In addition, the finding of the seasonal change of the CP branching ratio of K_L^0 meson will undoubtedly have some implications for quantum theory and special relativity. It will not only confirm the existence of preferred Lorentz frame, which may help to solve the problem of the incompatibility between quantum non-locality and special relativity, but also confirm the existence of the wavefunction collapse, which may help to settle the controversies about quantum theory and lead us to find a complete quantum theory. Certainly, it may also reveal the origin of CP violation, i.e., that CP violation results from the collapse of the wave function.

CHAPTER 10

Quantum Superluminal Communication

As we have shown, the collapse of the wave function requires the existence of a preferred Lorentz frame. This opens the first door to quantum superluminal communication (QSC). The possibility of realizing QSC will be detailedly analyzed in this chapter. We will demonstrate that the combination of the collapse of the wave function and the consciousness of the observer may permit the observer to distinguish nonorthogonal states in principle. This provides a principle for realizing QSC. A practical QSC scheme and some optimizing schemes are further proposed in terms of the QSC principle. Lastly, some implications of the existence of QSC are discussed.

10.1 A General Analysis

It is a plain fact that special relativity prohibits the existence of superluminal signalling. Their combination will result in the famous causal loop, which is forbidden in logic. However, the existence of quantum non-locality has indicated that the absolute validity of the principle of relativity may probably be limited. A solution to the problem of the incompatibility between quantum non-locality and special relativity is to assume the existence of a preferred Lorentz frame. In fact, it is widely argued that the property of quantum non-locality may actually require the existence of a preferred Lorentz frame. Moreover, the preferred Lorentz frame can also be detected through measuring the collapse time of the wave function according to the theory of quantum motion (cf. Chapter 9). If such a preferred Lorentz frame does exist, then QSC, which uses the quantum non-local influences to transfer information faster than light, will not result in the forbidden causal loop, and may exist (Gao 1999a; Suarez 2000b). This will open the first door to QSC.

Whereas quantum non-locality may defy special relativity and further permit the existence of QSC, it is very natural to guess that quantum non-locality may be used to realize QSC. However, the quantum theory also prohibits the existence of QSC. One of the efforts to realize QSC was made by Herbert (1982). He tried to decode the information contained in the quantum non-local influences by copying the state of a single particle. But Wootters and Zurek (1982) soon demonstrated that Herbert's copy method is forbidden by the (linear) quantum theory. They concluded that a single quantum cannot be cloned. In fact, there exists a more general demonstration proving that the existing quantum theory prevents the use of the quantum non-local influences for QSC. Eberhard (1978) and Ghirardi et al (1980) had given such demonstrations as early as the 1970s. One common conclusion within the framework of the existing quantum theory is that an unknown quantum state cannot be completely determined, and two arbitrarily given nonorthogonal states cannot be distinguished.

However, the above no-QSC conclusion may be not the last answer. For one reason, quantum theory is not the last theory. It is well known that the most serious problem in quantum theory is the measurement problem. The projection postulate doesn't tell us how and when the measurement result appears. In this sense, the existing quantum theory is an incomplete description of reality. A promising alternative to a complete quantum theory is the well-known revised quantum dynamics (cf. Chapter 3). In this theory, the instantaneous wavefunction collapse is replaced with a describable and dynamical collapse process. In fact, the existence of such dynamical processes is the common characteristic of a complete quantum theory. Our following analysis will only rely on this common characteristic.

Although no one has strictly demonstrated that revised quantum dynamics does not permit the existence of QSC, it is generally thought that the conclusion should be the same as that of the existing quantum theory. The reason is that these two theories give the same probability prediction about the measurement results. However, this conclusion doesn't consider all possible measurement situations. The usual no-QSC demonstration implicitly assumes that the observer does not intervene before the completion of the dynamical collapse of the wave function, and what the observer identifies is only the definite measurement result during a quantum measurement. In other words, it doesn't consider the unusual situation in which the observer (and especially his conscious perception) directly intervenes in the dynamical collapse process and may in fact exist in a quantum superposition. Since the dynamical collapse of the wave function is an objective process which is not related to the consciousness of the observer, the existence of the quantum superposition of the observer with consciousness can't be excluded in principle.

As we will see, the direct intervention of the conscious observer in quantum measurement may provide a possible way to realize QSC. Moreover, the realization of QSC is irrelevant to the concrete perception of the "quantum observer" in the superposition of definite perceptions.

10.2 The Principle of QSC

In this section, we will argue that the combination of the dynamical collapse of the wave function and the consciousness of the observer will permit the observer to distinguish nonorthogonal states in principle. This provides a principle for realizing QSC (cf. Gao 1999b, 2000, 2004a).

Let χ_1 and χ_2 be two different definite perception states of a conscious being, and $\chi_1 + \chi_2$ is the superposition state of such definite perception states. We assume the conscious being satisfies the following "QSC condition", i.e., that his perception time for the states χ_1 and χ_2, which is denoted by t_P, is shorter than the collapse time of the single superposition state $\chi_1 + \chi_2$, which is denoted by t_C, and that the time difference $\Delta t = t_C - t_P$ is large enough for him to identify. In the following, we will demonstrate that the conscious being can distinguish between the definite perception states χ_1 or χ_2 and the superposition state $\chi_1 + \chi_2$. This conclusion is irrelevant to the concrete collapse process of the superposition state $\chi_1 + \chi_2$.

First, we assume that a definite perception about the superposition state $\chi_1 + \chi_2$ can appear only after a dynamical collapse. This is a natural assumption, and accords with a basic scientific belief, i.e., that the conscious perceptions reflect the objective world as accurately as possible. Under this assumption, the conscious being can have a definite perception about the state χ_1 and χ_2 after the perception time t_P, but only after the collapse time t_C can the conscious being have a definite perception about the superposition state $\chi_1 + \chi_2$. When the conscious being satisfies the "QSC condition", he can be conscious of the time difference between t_P and t_C, then he can distinguish between the definite perception state χ_1 or χ_2 and the superposition state $\chi_1 + \chi_2$.

Secondly, we assume the conscious being can have a definite perception of the superposition state before the dynamical collapse has completed. We will demonstrate that the conscious being is also able to distinguish the states $\chi_1 + \chi_2$ and χ_1 or χ_2 with non-zero probability.

(1). If the definite perception of the conscious being in the superposed state $\chi_1 + \chi_2$ is neither χ_1 nor χ_2, then the conscious being can directly distinguish the states $\chi_1 + \chi_2$ and χ_1 or χ_2. For the latter, the definite perception of the conscious being is χ_1 or χ_2, but for the superposition state $\chi_1 + \chi_2$, the definite perception of the conscious being is neither χ_1 nor χ_2.

(2). If the definite perception of the conscious being in the superposed state $\chi_1 + \chi_2$ is χ_1, then the conscious being can directly distinguish the states $\chi_1 + \chi_2$ and χ_2. For the latter, the definite perception of the conscious being is χ_2, but for the superposition state $\chi_1 + \chi_2$, the definite perception of the conscious being is χ_1 before the collapse process finishes. In addition, the superposition state $\chi_1 + \chi_2$ will become χ_2 with probability 1/2 after the collapse process finishes. Then the definite perception of the conscious being will also change from χ_1 to χ_2 accordingly. For the state χ_1 or χ_2, the definite perception of the conscious being has no such change. Thus the conscious being is also able to distinguish the states $\chi_1 + \chi_2$ and χ_1 with probability 1/2.

(3). If the definite perception of the conscious being in the superposed state $\chi_1 + \chi_2$ is χ_2, the demonstration is similar to that of (2).

(4). If the definite perception of the conscious being in the superposed state $\chi_1 + \chi_2$ is random, i.e., one time it is χ_1, another time it is χ_2, then the conscious being can still distinguish the states $\chi_1 + \chi_2$ and χ_1 or χ_2 with non-zero probability. For the latter, the perception of the conscious being does not change. For the superposition state $\chi_1 + \chi_2$, the perception of the conscious being will change from χ_1 to χ_2 or from χ_2 to χ_1 with non-zero probability during the collapse process with independent randomness. For example, if the definite perception of the conscious being in the superposed state $\chi_1 + \chi_2$ is χ_1 before the collapse process finishes, but the superposition state becomes χ_2 after the collapse process finishes, then the perception of the conscious being will change from χ_1 to

χ_2. If the definite perception of the conscious being in the superposed state $\chi_1 + \chi_2$ assumes χ_1 or χ_2 with the same probability 1/2, then the above distinguishing probability will be 1/2.

In a word, we have demonstrated that if the conscious being satisfies the "QSC condition", he is able to distinguish the nonorthogonal states $\chi_1 + \chi_2$ and χ_1 or χ_2 with non-zero probability. This will directly lead to the availability of QSC[29]. We will give a practical scheme of QSC in the next section. It should be stressed that, since the collapse time of a single superposition state is an essentially stochastic variable, the "QSC condition" can be in principle satisfied in some collapse events with non-zero probability. For these stochastic collapse processes, the collapse time of the single superposition state is much longer than the (average) collapse time and the perception time of the conscious being. This provides an essential availability of QSC[30].

It is worth noting that the above conclusion is also valid in the many-worlds theory and Bohm's hidden variables theory. In the many-worlds theory (cf. Everett 1957; DeWitt and Graham 1973), the role of the decoherence process (cf. Guilini et al 1996; Zurek 1998) is similar to that of the dynamical collapse process in the revised quantum dynamics. Accordingly, the "QSC condition" for the conscious being will be that his perception time for the states χ_1 and χ_2 is shorter than the environment-induced decoherence time for the

---

[29] It should be noted that Squires (1992) had ever given an argument for superluminal signalling in the nonrelativistic domain. By comparison, the demonstration here is more consistent and complete. For a detailed comparison see Gao (2006d). In addition, Josephson and Pallikari-Viras (1991) also gave a general analysis of the biological utilisation of quantum non-locality.

[30] Note that some possible evidences (cf. Duane and Behrendt 1965; Targ and Puthoff 1974; Radin and Nelson 1989; Grinberg-Zylberbaum 1994; Wackermann 2003) have indicated that the human beings in some special states may satisfy the "QSC condition" and achieve QSC.

superposition state $\chi_1 + \chi_2$, and that the time difference is large enough for him to identify. Since the conscious perception process and the environment-induced decoherence process are essentially independent, the "QSC condition" can be satisfied in principle. Once the "QSC condition" is satisfied, the conscious being can also distinguish the nonorthogonal states $\chi_1 + \chi_2$ and χ_1 or χ_2 and further achieve QSC. In Bohm's hidden variables theory (cf. Bohm 1952), the conscious systems in the superposition state $\chi_1 + \chi_2$ and the definite state χ_1 or χ_2 will have different trajectories. Since the conscious system can be conscious of such difference, he can also distinguish the nonorthogonal states $\chi_1+\chi_2$ and χ_1 or χ_2 and achieve QSC. In fact, the realizability of QSC is irrelevant to the interpretations of quantum theory. It only bases on two firm facts: one is the existence of indefinite quantum superpositions, the other is the existence of definite conscious perceptions.

10.3 A Practical Scheme of QSC

In this section, we will give a practical scheme of achieving QSC based on the QSC principle. It includes two parts. The first part is how to distinguish nonorthogonal states. We design a device implementing this important function, which is called NSDD (Nonorthogonal States Distinguishing Device). The second part is how to achieve QSC using the hardcore device NSDD.

The implementation scheme of NSDD is as follows. The particles to be identified are photons, and the conscious being in the device can perceive a single photon[31]. Let the input states of the device be the nonorthogonal states $\psi_A+\psi_B$ or $\psi_A-\psi_B$ and ψ_A or ψ_B.

[31] In practical situation, a few photons may be needed.

ψ_A is the state of photon with a certain frequency entering into the eyes of the conscious being from the direction A, which can trigger a definite perception of the conscious being who perceives that the photon arrives from the direction A. ψ_B is the state of the photon with the same frequency entering into the eyes of the conscious being from the direction B, which can trigger a definite perception of the conscious being who perceives that the photon arrives from the direction B. $\psi_A+\psi_B$ and $\psi_A-\psi_B$ are the direction superposition states of the states ψ_A and ψ_B of photon. Let the initial perception state of the conscious being is χ_0, then after interaction the corresponding entangled state of the whole system is respectively $\psi_A \chi_A+\psi_B \chi_B$ and $\psi_A \chi_A-\psi_B \chi_B$ for the input states $\psi_A+\psi_B$ and $\psi_A-\psi_B$. The conscious being satisfies the "QSC condition", i.e., his perception time for the definite state $\psi_A \chi_A$ or $\psi_B \chi_B$, which is denoted by t_P, is shorter than the collapse time of the single superposition state $\psi_A \chi_A+\psi_B \chi_B$ or $\psi_A \chi_A-\psi_B \chi_B$, which is denoted by t_C, and that the time difference $\Delta t=t_C-t_P$ is large enough for him to identify[32]. According to the QSC principle, the device NSDD can distinguish the nonorthogonal states $\psi_A+\psi_B$ or $\psi_A-\psi_B$ and ψ_A or ψ_B. When the input state is ψ_A or ψ_B, the conscious being will perceive that the photon arrives from the direction A or B after the perception time t_p, and he assigns '1' as the output of the device[33]. When the input state is $\psi_A+\psi_B$ or $\psi_A-\psi_B$, the conscious being will perceive that the photon arrives from the direction A or B after the collapse time

[32] In an actual experiment, the conscious perceptions can be more accurately recorded by the EEG (electroencephalograph) recording of the observer, and the "QSC condition" can also be stated using the corresponding EEG recordings.

[33] In view of accuracy, an EEG may be used to record the perception time and produce the output of NSDD.

t_C, and he assigns '0' as the output of the device. NSDD can be implemented through the direct use of a conscious being or by an advanced consciousness simulation device in the future.

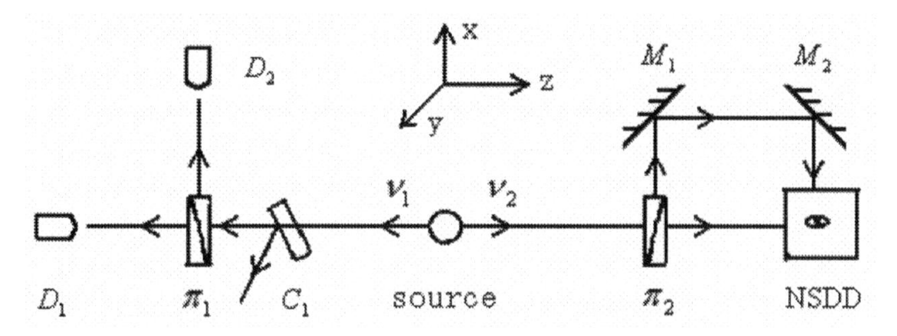

Figure 9 The setting of a practical QSC scheme

Now we will give the scheme of achieving QSC using the device NSDD. In reality, once the nonorthogonal photon states can be distinguished, achieving QSC will be an easy task, and it may be implemented by means of existing technology. Here we use the EPR polarization correlation pairs of photons as the carriers of information. We encode the outgoing information by operating the polarizer, and decode the incoming information using NSDD. The experimental setting is shown in the above figure. Pairs of photons, whose frequencies are ν_1 and ν_2, are emitted in the -z direction and +z direction from a source, are then analyzed by a one-channel polarizer π_1 and a two-channel polarizer π_2 respectively. The optical switch C_1 in the left side can be controlled to determine whether or not the photon ν_1 will pass to π_1. The transmission axes of the polarizers are both set in the direction *x*. The one-channel polarizer π_1 allows only the polarization components of the photon parallel to the transmission axis of the polarizer to be passed, and the two-channel polarizer π_2 allows the polarization components of the photon both parallel to and perpendicular to the transmission axis of the polarizer to be passed. The photon passed and analyzed by the polarizer π_1 is detected by D_1, and the photon analyzed by the two-channel polarizers π_2 is divided into two paths in space, and respectively input to NSDD from different directions.

We explain how QSC can be achieved by means of the above setting. Let the sender operate the optical switch C_1, and have the receiver observe the output of NSDD. Suppose the communication rules are stated as follows. The encoding rule for the sender is that not measuring the photon represents sending the code '0', and measuring the photon represents sending the code '1'[34]. The decoding rule for the receiver is that the output of NSDD being '0' represents having received the code '0', and the output of NSDD being '1' represents having received the code '1'.

The communication process can be stated as follows. When the sender wants to send a code '0', he controls the optical switch C_1 to let the photon ν_1 move freely and not be analyzed by the polarizer π_1. Then the state of the photon ν_2 is a superposition state like $\psi_A+\psi_B$ or $\psi_A-\psi_B$ after it passes the polarizer π_2, and the output of NSDD is '0'. The receiver can decode the sent code as '0'. When the sender wants to send a code '1', he controls the optical switch C_1 to allow the photon ν_1 to be analyzed by the polarizer π_1 and detected by D_1 before the photon ν_2 arrives at NSDD. Then the state of the photon ν_2 collapses to a definite state like ψ_A or ψ_B, and the output of NSDD is '1'. The receiver can decode the sent code as '1'. Thus the sender and receiver can achieve QSC using the above setting and communication rules.

---

[34] In practical situation, in view of the stochastic property of the collapse time and other possible errors, redundancy coding is required. A single information code should be encoded through the same operation on a small number of photons.

Since it may be very difficult for the conscious being to perceive a single photon, the superposition state of a small number of photons such as $|\varphi_1\varphi_2...\varphi_n>_{\nu_1}|\phi_1\phi_2...\phi_n>_{\nu_2}+|\phi_1\phi_2...\phi_n>_{\nu_1}|\varphi_1\varphi_2...\varphi_n>_{\nu_2}$ is needed to achieve QSC in a practical situation. Unfortunately it is also very difficult to achieve such a superposition state of many photons using existing technology. Here we will present an optimizing scheme. The method is to use a large number of entangled states of pair photons[35]. We assume pair photons are independently emitted from the source one after the other in the above experiment, and the other settings are the same. The state of the i-th pair photon is $|\varphi_i>_{\nu_1}|\phi_i>_{\nu_2}+|\phi_i>_{\nu_1}|\varphi_i>_{\nu_2}$. Then the state of many such independent pair photons will be $\prod_{i=1}^{n}(|\varphi_i>_{\nu_1}|\phi_i>_{\nu_2}+|\phi_i>_{\nu_1}|\varphi_i>_{\nu_2})$. The conscious being can still distinguish the nonorthogonal states $\prod_{i=1}^{n}(|\varphi_i>_{\nu_1}|\phi_i>_{\nu_2}+|\phi_i>_{\nu_1}|\varphi_i>_{\nu_2})$ and one of its branch states such as $|\phi_1>_{\nu_2}|\varphi_2>_{\nu_2}...|\varphi_n>_{\nu_2}$ as he satisfies the "QSC condition". If the sender wants to send a code '1', he can still control the optical switch C_1 to let the photons ν_1 be analyzed by the polarizer π_1 and detected by D_1 before the photons ν_2 arrives at NSDD. The receiver will identify the input state of the photons ν_2 as a randomly collapsed definite state such as $|x_1y_2...y_n>_{\nu_2}$, and decode the sent code as '0'. Similarly, if the sender wants to send a code '0', he can still control the optical switch C_1 to let the photons ν_1 move freely and not be analyzed by the polarizer π_1. Then the conscious being will identify the

[35] Note that this method may also help to overcome the limitation resulting from the inefficiency of the photon detector to some extent. Here we only need to collapse the superposition states using the photon detector, and the concrete detection recordings are not needed.

input state of the photons ν_2 as a superposition state $\prod_{i=1}^{n}(|x_i>_{\nu_1}|y_i>_{\nu_2}+|y_i>_{\nu_1}|x_i>_{\nu_2})$, and decode the sent code as '0'. Thus QSC can also be achieved using the above method. Evidently this experiment could more easily be conducted using existing technology, and may be completed in the near future.

It can be seen that the communication rate of QSC will be limited by the perception time of the conscious being[36], and this may prevent QSC from being widely applied. One optimizing scheme would be to combine QSC and quantum teleportation. QSC would be used to replace the classical communication required by quantum teleportation. Since the information transferred through this channel is very little, and the majority of information is transferred through the quantum channel in quantum teleportation, this combination will largely increase the communication rate of QSC. It is anticipated that advanced perception simulation technology may be available in the near future, and thus the communication rate of QSC would be largely increased.

QSC will have many advantages over conventional communication (cf. Gao 2001c). First, the transfer delay of QSC is irrelevant to the communication distance, and can be zero in principle. Thus QSC is the fastest communication method. Secondly, the carriers of information may not pass through the space between the sender and the receiver for QSC, thus the communication process is not influenced by the environment between them. Thus QSC is a kind of complete anti-jamming communication method. Thirdly, since the carriers of information can be stored only in the sender and receiver for QSC, a third party cannot eavesdrop the transferred information. Thus QSC is the most secure communication method.

---

[36] The perception time of human being is in the order of $0.1s$, thus the corresponding communication rate of QSC will be in the order of 10bps.

Lastly, as there is no electromagnetic radiation involved in QSC, it is a "green" communication method.

10.4 Further Discussions

In order to further understand the realization way of QSC, we will give a brief analysis of the relation between quantum collapse and consciousness, and discuss some implications of the existence of QSC.

Bohr (1927) first stressed the special role of measurement in quantum theory with his complementarity principle. Later von Neumann (1955) rigorously formulated the measurement process mathematically by means of the projection postulate, but the inherent vagueness in the definition of a measurement still exists. In order to explain how a definite result is generated by the measurement of an indefinite quantum superposition, the consciousness of the observer was invoked by von Neumann (1955). This theory was further advocated by Wigner (1967), according to which consciousness can break the linear superposition law of quantum mechanics. This may be the first statement made about the relationship between consciousness and quantum collapse. It states that consciousness results in the collapse of the wave function.

However, this relationship between quantum collapse and consciousness needs to be greatly revised when faced with the problem of quantum cosmology (cf. DeWitt 1967; Hartle and Hawking 1983). For the state of the whole universe, no outside measuring apparatus or observer exists. Thus the special role of measurement or observation is essentially deprived, and the collapse process, if it exists, must be added to the wave function. The dynamical collapse theory further revised the above relationship (cf. Chapter 3). In the dynamical collapse theory the normal linear evolution and the instantaneous collapse of the wave function are unified in a stochastic nonlinear Schrödinger equation, and the quantum collapse is a natural result of such evolution. Thus the new relationship between consciousness and quantum collapse is that the collapse of the wave function must happen independent of consciousness.

Although the collapse of the wave function does not need to resort to the consciousness of an observer, their combination may lead to some new results such as the availability of QSC. As the seeds of QSC, consciousness and quantum collapse are both indispensable. Quantum collapse provides the basis, and consciousness provides the means. Even if consciousness doesn't intervene, quantum collapse itself can also display quantum non-locality, and thus results in the existence of a preferred Lorentz frame. However, quantum collapse alone can't provide the means of realizing QSC, and its inherent randomness ruthlessly block the way. Here consciousness becomes a delicate bridge to QSC. The direct intervention of consciousness can help to obtain more information about the measured quantum state, which is enough to distinguish nonorthogonal states and decode the veiled information non-locally transferred by quantum collapse. This provides a way to realize QSC.

Furthermore, it can be seen that the distinguishability of nonorthogonal states will result in the violation of the quantum superposition principle. This indicates that consciousness will introduce a new kind of nonlinearity to the complete evolution of the wave function. The new non-linearity is definite, not stochastic. As we know, nonlinear quantum theory generally has some internal difficulties (cf. Gisin 1990; Polchinski 1991; Weinberg 1992; Czachor 1995). For example, the description of composite systems depends on a particular basis in a Hilbert space. However, as we have demonstrated, the consciousness of the observer will naturally select a privileged basis in its state space. Thus the nonlinear evolution introduced by consciousness is logically consistent and may exist (cf. Czachor 1995). On the other hand, once there exist nonlinear evolution of the wave function and real quantum collapse process, QSC must exist (cf. Gisin 1990; Czachor 1995). The reason is that nonlinear evolution doesn't conserve scalar products. States that are initially orthogonal will lose their orthogonality during the evolution. This is consistent with the above realization way of QSC.

Certainly, once QSC can be realized, we can directly detect the preferred Lorentz frame using the QSC process. In addition, QSC can also be used as a natural method to synchronize the clocks in different positions, and the simultaneity can be uniquely defined using such superluminal signal (cf. Gao 2004c). As a result, the one-way speed of light can be measured. Based on the superluminal synchrony method, the space-time transformation will be not the usual Lorentz transformation, but a new superluminal space-time transformation, which holds the absoluteness of simultaneity (cf. Gao 2004c). This may have some implications for the space-time structure and the final unification of quantum and gravity.

Lastly, we note that the QSC principle may provide a physical method for distinguishing between man and machine (cf. Gao 2002b, 2004b). We can test whether the conner possesses consciousness through its identification of nonorthogonal states. The conner with consciousness can distinguish nonorthogonal states, whereas the conner without consciousness cannot. Moreover, the QSC principle may further imply that consciousness is not reducible or emergent, but a new fundamental property of matter (cf. Gao 2003a, 2006c, 2006e). It is expected that a complete theory of matter must describe all properties of matter, thus consciousness, the new fundamental property of matter, must enter the theory from the start. We will put forward a unified quantum theory of matter and consciousness in the next chapter.

CHAPTER 11

A Quantum Theory of Consciousness

Consciousness is the most familiar phenomenon. There are two distinct processes relating to the phenomenon: one is the objective matter process such as the neural process in the brain, the other is the concomitant subjective conscious experience. The relationship between matter and consciousness presents a well-known hard problem for science (cf. Chalmers 1996). It retriggers the debate about the long-standing dilemma of panpsychism versus emergentism recently (cf. Seager 1999, 2001). Panpsychism asserts that consciousness is a fundamental feature of the world that exists throughout the universe. Emergentism asserts that consciousness appears as an emerging result of the complex matter process. It is generally accepted that an essential separation of matter and consciousness will preclude any real integration of consciousness with the present scientific picture of the physical world, and panpsychism and emergentism are the two main positions that can complete the integration. Then we must decide whether consciousness emerges from mere matter or whether consciousness is a fundamental property of matter.

Emergentism is the most popular solution to the hard problem of consciousness. But many doubt that it can bridge the explanation gap ultimately (cf. Chalmers 1996). On the other hand, although panpsychism may provide an attracting and promising way to solve the hard problem, it also encounters some serious problems. It is widely argued that the physical world is causally closed, and the consciousness property assigned by panpsychism must lack all causal efficacies, i.e., there is a purely physical explanation for the occurrence of every physical event and the explanation does not refer to any consciousness property (cf. McGinn 1999). But if panpsychism is true, the consciousness property should take part in the causal chains of the physical world and should present itself in our investigation of the physical world. Then whether or not do the causal efficacies of consciousness exist? If they do exist, how to find them?

In this chapter we will try to solve the above problems by studying the combination of quantum and consciousness. We will show that the consciousness of the observer can help to distinguish between the definite states and the quantum superposition of them as the QSC principle reveals, while the usual physical measuring apparatus without consciousness cannot. This result indicates that the causal efficacies of consciousness do exist when considering the basic quantum processes, and thus consciousness is not reducible or emergent, but a new fundamental property of matter. On the basis of the new analysis, we will finally present a unified quantum theory of matter and consciousness.

11.1 Consciousness and Physical Measurement

We will first analyze the role of consciousness in physical measurement (cf. Gao 2004b). Physical measurement generally consists of two processes: (1). the physical interaction between the observed object and measuring apparatus; (2). the psychophysical interaction between the measuring apparatus and the observer. In some special situations, measurement may be the direct interaction between the observed object and the observer.

Even though what physics commonly studies are the insensible objects, the consciousness of the observer must take part in the last phase of measurement. The observer is introspectively aware of his perception of the measurement result. Here consciousness is used to end the infinite chains of measurement. This is one of the main differences between the functions of a physical measuring apparatus and an observer in the measurement process. But unfortunately the difference seems to be not identifiable using physical methods. Then whether or not does the consciousness of the observer have some physically identifiable effects that are lacking for the physical measuring apparatus?

In classical theory, the influence of the measuring apparatus or the observer on the observed object can be omitted in principle during a measurement, and the psychophysical interaction between the observer and the measuring apparatus does not influence the reading of the pointer of the measuring apparatus either. Thus measurement is only an ordinary one-to-one mapping from the state of the observed object to the pointer state of the measuring apparatus and then to the perception state of the observer, or a direct one-to-one mapping from the state of the observed object to the perception state of the observer. The consciousness of the observer has no physically identifiable functions that are different from those of the physical measuring apparatus in classical theory.

In quantum theory, however, the influence of the measuring apparatus on the observed object cannot be omitted owing to the existence of quantum superposition. It is just this influence that generates the definite measurement result to some extent. Since the measuring apparatus has generated one definite measurement result, the psychophysical interaction between the observer and the measuring apparatus is still an ordinary one-to-one mapping, and the process is the same as that in classical situation. But when the observed object and the observer directly interact, quantum evolution will introduce a new element to the psychophysical interaction between the observer and the measured object. It will lead to the appearance of the conscious observer in quantum superposition. In the following section, we will show that the consciousness of the observer in superposition state can indeed have some physically identifiable effects that are lacking for the physical measuring apparatus.

11.2 A Quantum Effect of Consciousness

As we know, the usual measurement using the physical measuring apparatus cannot distinguish nonorthogonal states. But when the physical measuring apparatus is replaced by a conscious being and considering the influence of consciousness, nonorthogonal states can be distinguished in principle according to the QSC principle (cf. Chapter 10). Accordingly the distinguishability of nonorthogonal states reveals a quantum effect of consciousness, which is lacking for the physical measuring apparatus.

We illustrate the quantum effect of consciousness with a black box system. We first define a rule, i.e., that the outputs of the system are respectively '0' and '1' for the inputs ψ_1 and ψ_2, and the outputs of the system are '2' for other inputs.

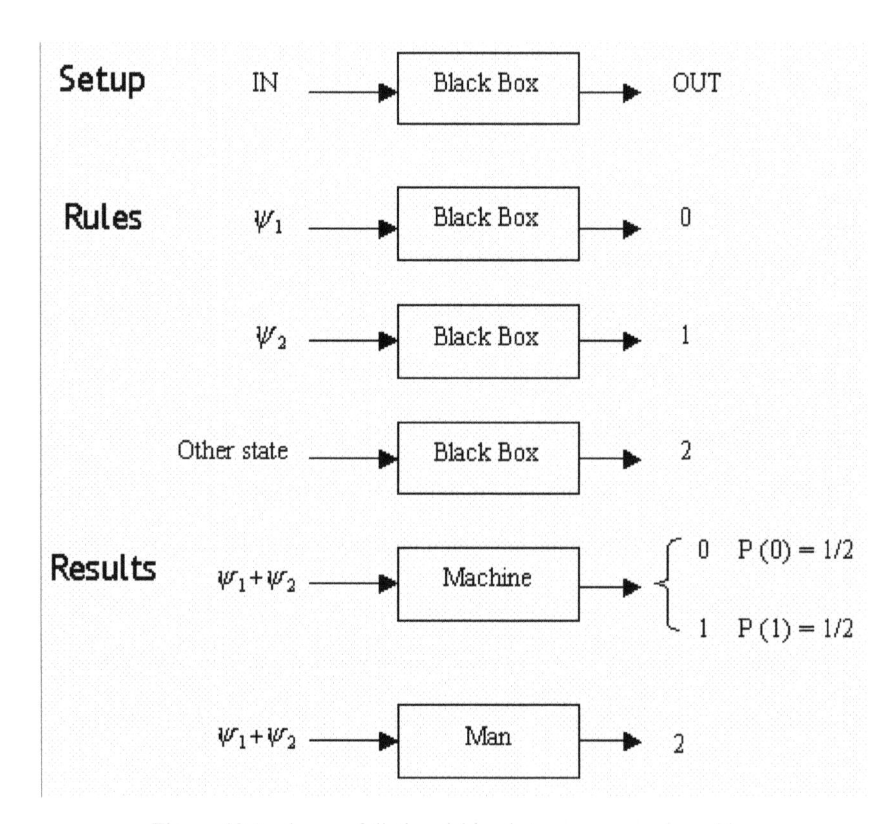

Figure 10 A scheme of distinguishing between man and machine

Now input the superposition state $\psi_1 + \psi_2$ to the system. If a machine without consciousness is in the black box, the output will be '0' or '1' with probability 1/2. The machine without consciousness cannot distinguish the nonorthogonal states $\psi_1 + \psi_2$ and

ψ_1 or ψ_2. As a result, it identifies the superposition state $\psi_1+\psi_2$ with the state ψ_1 or ψ_2. Then the output will be '0' or '1' according to the rule. However, if a conscious being is in the black box, the output will be '2'. The conscious being can distinguish the nonorthogonal states $\psi_1+\psi_2$ and ψ_1 or ψ_2. As a result, he regards the superposition state $\psi_1+\psi_2$ as a state different from ψ_1 and ψ_2. Then the output will be '2' according to the rule. The different outputs for man and machine clearly reveal a quantum effect of consciousness, which is lacking for the machine without consciousness. Certainly, such an effect can also be used to differentiate between man and machine (cf. Gao 2000, 2002b, 2004b).

11.3 Consciousness Is a Fundamental Property of Matter

In this section, we will further demonstrate that consciousness is not reducible or emergent, but a fundamental property of matter. This may provide a quantum basis for panpsychism.

As we have shown, the conscious being or the matter with consciousness can distinguish nonorthogonal states, while the usual physical measuring apparatus or the matter without consciousness cannot. Here we note that consciousness contains two aspects: one is the conscious property, the other is the conscious content. The former denotes the ability of being conscious of something, and can be described by the conscious strength. The latter denotes the content being consciously perceived by the subject. The conscious content can be described by its complexity. In order to distinguish nonorthogonal states, it is required that the measuring system possesses the conscious property, and its conscious content at least contains the perceptions of the nonorthogonal states.

If consciousness is reducible or emergent, then the matter with consciousness should also follow the basic physical principles such as the principle of energy conservation etc. According to the quantum principles, nonorthogonal states cannot be distinguished. However, as we have shown, the conscious being or the matter with consciousness can distinguish nonorthogonal states. This clearly indicates that consciousness violates the quantum principles, which are basic physical principles. Thus the consciousness property should be not reducible or emergent, but a new fundamental property of matter[37]. It should be not only possessed by the conscious being, but also possessed by the atoms as well as the physical measuring apparatus. The difference only lies in the conscious content. The existing experience shows that the conscious content of a human being can be very complex, while the conscious content of a physical measuring apparatus, if it exists, may be very simple. Such simple conscious content cannot help to distinguish nonorthogonal states.

On the other hand, if consciousness is a new fundamental property of matter, then it is very natural that it violates the existing basic physical principles, in which consciousness is not included as one fundamental property of matter. It is expected that a complete theory of matter must describe all properties of matter, thus consciousness, the new fundamental property of matter, must enter the theory from the start. Since the distinguishability of nonorthogonal states violates the basic linear superposition principle of quantum theory, the consciousness property of matter will introduce a new nonlinear evolution term to the complete equation of the wave function when the conscious content is complex enough. The nonlinearity is not stochastic, but definite. It has been argued that the nonlinear quantum evolution introduced by consciousness is logically consistent and may exist (cf. Czachor 1995;

[37] Note that if the wavefunction collapse results from the consciousness of the observer (cf. von Neumann 1955; Wigner 1967; Stapp 1993), then collapsing the wave function will be a quantum effect of consciousness, and thus consciousness should be also a fundamental property of matter. In addition, we stress again that the above conclusion is independent of the interpretations of quantum theory. It only depends on two firm facts: one is the existence of indefinite quantum superpositions, the other is the existence of definite conscious perceptions.

Gao 2004a). In addition, we may use the above nonlinear term in the complete evolution equation of matter to define the consciousness property of matter. Then just like the other properties of matter such as mass and charge etc, consciousness is also a fundamental property of matter that can be strictly described in mathematics to some extent.

The above argument provides a quantum basis for panpsychism (cf. Gao 2003a, 2006c, 2006e). As we know, one of the most severe problems of panpsychism is the apparent lack of evidence that the fundamental entities of the physical world such as electrons and protons possess any consciousness features. Certainly, such "no evidence" argument can be reasonably disputed by noting that there may not exist any signs of complex consciousness at the simplest level (e.g. the conscious content is very simple or even null), and it may be very difficult to see them even when they do exist there. The existence of gravitation is a good example. Its extreme weakness between the fundamental entities does not disconfirm that gravitation is not a fundamental feature of the physical world (cf. Seager 1999, 2001). Now the existence of the definite nonlinear evolution introduced by consciousness may further help to solve the above problem. Since such definite nonlinearity can be experimentally tested even for the evolution of the fundamental entities such as electrons and protons, it may provide a well-grounded and promising way to confirm the panpsychism doctrine by experiment.

11.4 Conscious Process as Quantum Computation

Whereas consciousness is a fundamental property of matter, it may be expected that its evolution is essentially a quantum process. In this section, we will argue that the conscious process is a kind of quantum computation, and the definite conscious experience appears as the result of such quantum computation (cf. Gao 2006c).

Everyday experience shows that a definite conscious experience can only be obtained through a process finished in finite time interval, while the process itself is unconscious. The existence of finite conscious time is also confirmed by experiment (cf. Libet 1993). Whereas consciousness is a fundamental property of matter, the existence of such transition process from pre-consciousness to consciousness may imply that the conscious process is essentially a quantum collapse process. If the conscious process is a classical process, then since consciousness is a fundamental property of matter, which is not generated by the matter process, the system with consciousness should be conscious of its state and process at all times. As a result, the transition process from pre-consciousness to consciousness will not exist. On the other hand, if the conscious process is a quantum collapse process, then the collapse of a quantum superposition of definite conscious states into one of the definite conscious states will naturally correspond to the transition process from pre-consciousness to consciousness. Note that the system in a quantum superposition of definite conscious states does not possess a definite conscious experience. Thus we conclude that the conscious process may be a quantum collapse process. Whereas the information processing via quantum process is generally called quantum computation, the conscious process will be a kind of quantum computation.

In addition, some psi phenomena such as telepathy, if they do exist, may also imply that the conscious process involves quantum computation. Some experiments have primarily revealed that the information transfer between the human brains can be achieved in a certain non-local way (cf. Duane and Behrendt 1965; Targ and Puthoff 1974; Puthoff and Targ 1976; Radin and Nelson 1989; Grinberg-Zylberbaum et al 1994; Wackermann 2003). When the classical communication tunnels between the human brains are shut down, the information can also be transferred between them. This kind of information transfer process, if it really exists, will strongly imply that such process is a kind of quantum non-local process between two entangled brain systems, and the involved brain process or conscious process is a kind of quantum computation. On the other hand, the combination of quantum collapse and consciousness can indeed lead to the availability of non-local information transfer or superluminal communication between the conscious systems (cf. Chapter 10). This further supports the conclusion that the conscious process is a kind of quantum computation.

Quantum computation is a parallel computation process (cf. Nielsen and Chuang 2000). During such a process, the information, which distributes in different space regions spread by the state, not only combines to form a whole, but also is processed simultaneously. Thus if the conscious process is a kind of quantum computation, then the final result of such quantum computation will appear as a whole conscious experience with abundant binding content. This suggests that the combination problem and the binding problem may be more properly solved in the quantum framework (cf. Seager 1999). The whole formed by quantum entanglement is a kind of essentially inseparable quantum whole, each part of which does not possess independent conscious content as well as independent matter state. Only the whole system in the quantum entangled state can possess an inseparable conscious content. By comparison, the classical whole is a kind of essentially separable whole, the parts of which possess independent conscious content as well as independent matter states when they are in space-like separation. Thus the combination of classical parts cannot form a new inseparable whole, whereas the combination of quantum parts can form a new inseparable wholeness through quantum entanglement. This indicates that the combination problem can be more properly solved in the quantum framework. In addition, quantum entanglement can also provide a proper way to bind the conscious content distributing in different space regions, and help to solve the binding problem.

As a typical example, the conscious state of human brain is an inseparable wholeness, and its parts such as neurons do not possess their independent conscious states. This character essentially coincides with that of quantum entangled state. Furthermore, we can basically work out the conscious time or the collapse time of the superposition of different conscious perceptions using the theory of quantum motion. It will be shown that the theoretical value is quantitatively consistent with the measured value. This also supports the conclusion that the conscious process is a kind of quantum computation. As we know, the number of the neurons which can form a definite conscious perception is in the order of 10^4. In each neuron, the main difference between the activation state and the resting state lies in the motion of 10^6 sodium ions (Na^+) passing through the membrane. Since the membrane potential is in the order of 10^{-2} V, the energy difference between the activation state and the resting state is approximately 10^4 eV. According to the theory of quantum motion, the collapse time of the superposition of the activation state and the resting state of one neuron is

$$\tau_c \approx \frac{\hbar E_p}{(\Delta E)^2} \approx (\frac{2.8Mev}{0.01MeV})^2 \approx 10^5 s \qquad (11.1)$$

where $\hbar$ is the Planck constant divided by 2π, E_p is the Planck energy, and ΔE is the energy difference between the states in the superposition. Thus the collapse time of the superposition of two different conscious perceptions is

$$\tau_c \approx (\frac{2.8Mev}{100MeV})^2 \approx 1ms \qquad (11.2)$$

In this superposition state, one conscious perception state approximately contains 10^4 neurons in the activation state, and the other conscious perception state approximately contains 10^4 neurons in the resting state. This result shows that the theoretical value of the collapse time of the superposition state of different conscious perceptions or the conscious time is in the order of several milliseconds. On the other hand, the measured value of the conscious time of human brain is in the order of several hundred milliseconds (cf. Libet 1993). Whereas a complex conscious process may generally contain many successional collapse processes, these two values are very consistent. In addition, the above theoretical value of the conscious time also coincides with the coherent 40Hz oscillation of neurons accompanying the appearance of conscious experience (cf. Crick 1994).

It is worth noting that some concrete models of consciousness such as the Orch OR model also suggest that the conscious process is a kind of quantum computation. In the Orch OR model, the protein assemblies called microtubules within the brain's neurons are viewed as self-organizing quantum computers (cf. Penrose 1994; Hameroff and Penrose 1996). It is generally argued that the brain's neurons seem unsuitably warm and wet for delicate quantum computation which would be susceptible to thermal noise and environmental decoherence (cf. Tegmark 2000). However, a recent calculation suggests that microtubules can avoid environmental decoherence long enough to achieve quantum computation (cf. Hagan, Hameroff and Tuszynski 2002).

11.5 A Unified Theory of Matter and Consciousness

Since consciousness is a fundamental property of matter, the complete matter state should include the conscious content. Accordingly a unified theory of matter and consciousness should include two parts: one is the psychophysical principle, which states the connection between conscious content and matter state, the other is the complete quantum evolution law of matter state including conscious content. Such complete evolution includes three evolution terms: the first is the linear Schrödinger term, the second is the stochastic nonlinear term resulting in the dynamical collapse of the wave function, and the last is the definite nonlinear term introduced by consciousness (cf. Gao 2006c).

Undoubtedly it is very difficult to find the psychophysical principle. Some important analyses have been presented (cf. Chalmers 1996). Here we mainly discuss the definite nonlinear evolution term introduced by consciousness. Although its final form has not been found, we may give a primary analysis of its characters. As we have shown, the definite nonlinear evolution appears in the following quantum process:

$$(\psi_1+\psi_2)\chi_0 \rightarrow (\psi_1\chi_1+\psi_2\chi_2)\chi_{12} \tag{11.3}$$

where χ_{12} denotes the state in which the conscious being perceives that the observed state is a superposition state, not a definite state. The appearance of χ_{12} indicates that the evolution is nonlinear. It also shows that consciousness results in a special change of matter state during the nonlinear evolution, which cannot brought by the usual properties of matter. Since the change of matter state generally corresponds to the change of energy distribution among the parts of the system, the definite nonlinear evolution introduced by consciousness will change the energy distribution among the parts of the system. As we have argued, the conscious process essentially involves quantum computation, and the conscious system is generally in a quantum entangled state. Thus the definite nonlinear evolution introduced by consciousness can change the energy distribution among the parts of the entangled system. Owing to the non-local character of quantum entanglement, the evolution may also change the energy distribution among the parts of a larger entangled system including the conscious system and other external systems. This analysis implies that the definite nonlinear term introduced by consciousness may have some kind of fundamental form, and the corresponding evolution can also bring some more basic effects.

As an example, we will predict a new quantum effect of consciousness resulting from the definite nonlinear evolution introduced by consciousness. Since the definite nonlinear evolution does not preserve the orthogonality of the states, such evolution can change the coherence of the branches of the state of an external system entangled with the conscious system and further change the statistic behavior of the external system. As a result, the definite nonlinear evolution introduced by consciousness may influence the statistic distribution of the measurement results of an external random process, and there may also exist a correlation between the influenced results and the conscious content. We note that some experiments (cf. Radin and Nelson 1989; Jahn et al 1997; Ibison and Jeffers 1998; Jeffers 2003) may have primarily revealed such a quantum effect of consciousness.

The above analysis presents a basic framework for a unified theory of matter and consciousness. This unified theory will not only tell us how the state of matter with consciousness evolves, but also tell us how the conscious content relates to the matter state. As a prediction of the theory, since consciousness is a fundamental property of matter, and there exists a connection between the conscious content and the matter state, a conscious machine can be constructed. It can be reasonably guessed that the simplest conscious machine which can distinguish two given nonorthogonal states may be only composed of several qubits. Certainly, in order to build up a complete theory of matter and consciousness, we need the organic combination of quantum theory, information science, neuroscience, cognitive science and psychology etc. This may be the biggest challenge to science in the 21st century.

11.6 Some Suggested Experiments

In order to confirm the existence of the quantum effects of consciousness, which is the core of the demonstrations in this chapter, we propose the following experimental schemes. The experiments can be conducted using human beings or animals.

1. Control experiment

Produce several photons with a certain frequency. Input them to the eyes of the subject. Test and record the conscious time of the subject through EEG (electroencephalograph) or his dictation.

2. Quantum perception experiment I

Produce a direction superposition state of the photons with the same frequency[38]. Input one branch of the superposition state to the eyes of the subject, and let the other branch freely spread (not input to a measuring apparatus). Test whether the subject perceives the photons during the normal conscious time.

3. Quantum perception experiment II

Produce a direction superposition state of the photons with the same frequency. Input both branches of the superposition state to the eyes of the subject. Test whether the subject perceives the photons during the normal conscious time.

4. Perceptions entanglement experiment I

Produce a direction superposition state of the photons with the same frequency. Input the two branches of the superposition state to the eyes of two independent subjects respectively. Test whether the subjects perceive the photons during the normal conscious time. It is suggested that the subjects are unfamiliar with each other before the experiment. This can be further validated by the phase incoherence of their brain waves.

If the subjects perceive the photons after a time interval longer than their normal conscious time in any case of the above experiments, then we will have confirmed the existence of the unusual "QSC condition" in human brains, which can result in the quantum effects of consciousness. In addition, it is suggested that the subjects in the above experiments should be composed of three independent groups at least. The subjects in the first group are in normal conscious state, the subjects in the second group are in meditation state, and the subjects in the third group are in qigong state.

---

[38] We may also use the other kinds of superposition state of the photons such as the polarization superposition state.

5. Perceptions entanglement experiment II

Produce a direction superposition state of the photons with the same frequency. Input the two branches of the superposition state to the eyes of two independent subjects isolated in space respectively. Then stimulate one of the subjects using flashes or visual patterns at random intervals. Record his evoked potentials and the corresponding brain activities of the other subject. Test whether there exists a statistical relevance between them. At the same time, ask the subjects whether they have the conscious perceptions relating to the stimulations. The appearance of the statistical relevance may confirm the existence of the quantum effects of consciousness.

References

Adler R. J. and Santiago D. I. (1999) Mod. Phys. Lett. A 14, 1371.

Adler S. L. and Horwitz L. P. (1999) preprint quant-ph/9909026.

Adler S. L. (2001) preprint quant-ph/0109029.

Adler S. L. and Brun T. A. (2001) J. Phys. A 34, 4797-4809.

Aharonov Y. and Vaidman L. (1993) Phys. Lett. A 178, 38-42.

Aharonov Y., Anandan J. and Vaidman L. (1993) Phys. Rev. A 47, 4616-4626.

Albert D. (2000) *Time and Chance.* Cambridge: Harvard University Press.

Allen S. W. et al (2004) Mon. Not. Roy. Astron. Soc. 353, 457.

Amelino-Camelia G. (2002) Nature 418, 34.

Arntzenius F. (2000) The Monist 83(2), 187-208.

Arntzenius F. (2003) Studies in the History and Philosophy of Modern Physics 34B, 281-282.

Arntzenius F. (2004) PhilSci Archive 1792.

Bekenstein J. D. (1981) Phys. Rev. D 23, 287.

Bell J. S. (1964) Physics 1, 195.

Bell J. S. (1986) in *The Ghost In the Atom*, Davies P. and Brown J. eds. Cambridge: Cambridge University Press.

Bell J. S. (1987) *Speakable and Unspeakable in Quantum Mechanics*. Cambridge: Cambridge University Press.

Berg B.A. (1996) preprint hep-ph/9609232.

Bohm D. (1952) Phys. Rev 85, 166-193.

Bohr N. (1927) Nature 121, 580-590.

Bunge M. (1973) *Philosophy of Physics.* Dordrecht: Reidel.

Butterfield J. and Isham C. J. (2001) in *Physics meets Philosophy at the Planck Scale.* Callender C. and Huggett N. eds. Cambridge: Cambridge University Press.

Butterfield J. (2005) PhilSci Archive, 2553

Cao T. Y. (1997) *Conceptual Developments of Twentieth Century Field Theories.* Cambridge: Cambridge University Press.

Chalmers D. (1996) *The Conscious Mind.* Oxford: Oxford University Press.

Christian J. (2001) in *Physics meets Philosophy at the Planck Scale.* Callender C. and Huggett N. eds. Cambridge: Cambridge University Press.

Cohen A., Kaplan D. and Nelson A. (1999) Phys. Rev. Lett 82, 4971.

Cohn D. L. (1993) *Measure Theory.* Boston: Birkhäuser.

Cova S., Ghioni M., Lacaita A., Samori C. and Zappa F. (1996) Appl. Optics 35, 1956.

Crick F. (1994) *The Astonishing Hypothesis.* New York: Scribner's.

Czachor M. (1995) preprint quant-ph/9501007.

d'Espagnat B. (2003) *Veiled Reality.* Massachusetts: Addison-Wesley. 2d. ed.

d'Espagnat B. (2006) *On Physics and Philosophy.* Princeton: Princeton University Press.

Deutsch D. (1985) Int. J. Theo. Phys 24, 1-41.

DeWitt B. S. (1967) Phys. Rev 160, 1113.

DeWitt B. S. and Graham N. eds. (1973) *The Many-Worlds Interpretation of Quantum Mechanics.* Princeton: Princeton University Press.

Diosi L. (1989) Phys. Rev. A 404, 1165.

Duane T. D. and Behrendt T. (1965) Science 150, 367.

Eberhard P. H. (1978) Nuovo Cimento B 46, 392.

Eidelman S. et al. (Particle Data Group) (2004) Phys. Lett. B 592, 1. (URL: http://pdg.lbl.gov)

Einstein A., Podolsky B. and Rosen N. (1935) Phys Rev 47, 777-780.

Everett H. (1957) Rev. Mod. Phys 29, 454-462.

Feynman R. P., Leighton R. B. and Sands M. (1963) *The Feynman Lecture on Physics.* Vol III. Massachusetts: Addison-Wesley.

Feynman R. P. (1995) *Feynman Lectures on Gravitation.* Hatfield B. eds. Massachusetts: Addison-Wesley.

Fivel D. I. (1997a) Phys. Rev. A 56, 146-156.

Fivel D. I. (1997b) preprint quant-ph/9710042.

Floyd S. (2003) UCSD Graduate Philosophy Conference draft.

Ford L. H. (1997) preprint gr-qc/9707062.

Gao S. (1993) Unpublished manuscript.

Gao S. (1999a) preprint quant-ph/9906113.

Gao S. (1999b) preprint quant-ph/9906116.

Gao S. (1999c) preprint physics/9907001, physics/9907002.

Gao S. (2000) *Quantum Motion and Superluminal Communication.* Beijing: Chinese Broadcasting & Television Publishing House. (in Chinese)

Gao S. (2001a) Physics Essays 14 (1), 37-48.

Gao S. (2001b) PhilSci Archive, 453.

Gao S. (2001c) Chinese Patents Gazette 17(41), 122.

Gao S. (2002a) preprint quant-ph/0209022.

Gao S. (2002b) The Noetic Journal 3(3), 233-235.

Gao S. (2003a) NeuroQuantology 1(1), 4-9.

Gao S. (2003b) *Quantum.* Beijing: Tsinghua University Press. (in Chinese)

Gao S. (2004a) Found. Phys. Lett 17(2), 167-182.

Gao S. (2004b) Axiomathes: An International Journal in Ontology and Cognitive Systems 14 (4), 295-305.

Gao S. (2004c) Bulletin of Pure and Applied Sciences Series D (Physics) 23 (2), 75-80.

Gao S. (2005) Chin. Phys. Lett 22, 783-784.

Gao S. (2006a) Galilean Electrodynamics 17(1), 3-10.

Gao S. (2006b) Int. J. Theo. Phys. May 26 (Online First).

Gao S. (2006c) NeuroQuantology 4(2), 166-185.

Gao S. (2006d) Infinite Energy 68, 17-21.

Gao S. (2006e) in *The Handbook of Whiteheadian Process Thought*. Weber M. eds. Frankfurt: Ontos Verlag.

Garay L. J. (1995) Int. J. Mod. Phys A 10, 145.

Ghirardi G. C., Rimini A. and Weber T. (1980) Letters Nuovo Cimento 27, 293.

Ghirardi G. C., Rimini A. and Weber T. (1986) Phys. Rev. D 34, 470.

Ghirardi G. C., Pearle P. and Rimini A. (1990) Phys. Rev. A 42, 78.

Gisin N. (1990) Phys. Lett. A 143, 1-2.

Gong Y. G., Wang B. and Zhang Y. Z. (2004) preprint hep-th/0412218.

Grinberg-Zylberbaum J., Dalaflor D., Attie L. and Goswami A. (1994) Physics Essays 7, 422.

Guilini D., Joos E., Kiefer C., Kupsch J., Stamatiscu I. O. and Zeh H. D. (1996) *Decoherence and the Appearance of a Classical World in Quantum Theory*. Berlin: Springer-Verlag.

Hagan S., Hameroff S. R. and Tuszynski J. A. (2002) Phys. Rev. D 65, 061901.

Hameroff S. R. and Penrose R. (1996) Journal of Consciousness Studies 3 (1), 36-53.

Hardy L. (1992) Phys. Rev. Lett 68, 2981-2984.

Hartle J. B. and Hawking S. W. (1983) Phys. Rev 28, 2960.

Herbert N. (1982) Found. Phys 12, 1171.

Hooft G. 't (1993) preprint gr-qc/9310026.

Horava P. and Minic D. (2000) Phys.Rev.Lett. 85, 1610.

Hughston L.P. (1996) Proc. Roy. Soc. Lond. A 452, 953.

Huterer D. and Cooray A. (2004) preprint astro-ph/0404062.

Ibison M. and Jeffers S. (1998) J. Scientific Exploration 12(4), 543-550.

Jahn R. G., Dunne B. J., Nelson R. D., Dobyns Y. H. and Bradish G. J. (1997) J. Scientific Exploration 11(3), 345-367.

Jeffers S. (2003) Journal of Consciousness Studies 10(6-7).

Josephson B. D. and Pallikari-Viras F. (1991) Found. Phys 21, 197-207.

Kleinknecht K. (2003) *Uncovering CP Violation.* Springer tracts in modern physics 195.

Lewis D. (2001) Philosophy and Phenomenological Research 63, 381-398.

Libet B. (1993) *Neurophysiology of Consciousness: Selected Papers and New Essays.* Boston: Birkhauser.

Marshall W., Simon C., Penrose R. and Bouwmeester D. (2003) Phys. Rev. Lett 91, 130401.

McGinn C. (1999) *The Mysterious Flame: Conscious Minds in a Material World.* New York: Basic Books.

Morgan J. C. II (1989) *Point set Theory.* New York: Marcel Dekker.

Nielsen M. A. and Chuang I. L. (2000) *Quantum Computation and Quantum Information*, Cambridge: Cambridge University Press.

Nielsen O. A. (1994) *An Introduction to Integration and Measure Theory.* Kingston: John Wiley and Sons.

Papa-Grimaldi Alba. (1996) The Review of Metaphysics 50, 299-314.

Pearle P. (1989) Phys. Rev. A 39, 2277.

Pearle P. (1999) in *Open Systems and Measurement in Relativistic Quantum Theory.* Petruccione F. and Breuer H. P. eds. New York: Springer Verlag.

Pearle P. (2004) Phys. Rev. A 69, 42106.

Penrose R. (1994) *Shadows of the Mind: a search for the missing science of consciousness.* Oxford: Oxford University Press.

Penrose R. (1996). Gen. Rel. and Grav. 28, 581-600.

Percival I. C. (1994) Proc. Roy. Soc. Lond. A, 447, 189-209.

Percival I. C. (1998) Phys.Lett.A. 244, 495-501.

Percival I. C. (2000) Proc. Roy. Soc. Lond. A 456, 25-37.

Perlmutter S. et al (1999) Astrophys. J. 517, 565.

Polchinski J. (1991) Phys. Rev. Lett. 66, 397.

Polchinski J. (1998) *String Theory.* Cambridge: Cambridge University Press.

Primas H. (1981) *Chemistry, Quantum mechanics and Reductionism.* Heidelberg: Springer.

Puthoff H. E. and Targ R. (1976) Proc. IEEE 64, 329-354.

Radin D. I. and Nelson R. D. (1989) Found. Phys. 19(12), 1499-1514.

Ratra B. and Peebles P. J. E. (1988) Phys. Rev. D 37, 3406.

Redhead P. A. (1996) *Macmillan Encyclopedia of Physics* Vol. IV. Rigden J. S. eds. New York: Simon & Schuster Macmillan.

Riess A. G. et al (1998) Astron. J. 116, 1009.

Riess A. G. et al (2004) Astron. J. 607, 665.

Rovelli C. (2000) preprint gr-qc/0006061.

Rugh S. E. and Zinkernagel H. (2002) Studies in History and Philosophy of Modern Physics 33, 663.

Russell B. (1903) *The Principles of Mathematics.* London: Allen and Unwin.

Salecker H. and Wigner E. P. (1958) Phys Rev 109, 571.

Salmon W. C. (eds). (1970) *Zeno's Paradoxes*. New York: Bobbs-Merrill.

Scarani V., Tittel W., Zbinden H. and Gisin N. (2000) Phys. Lett. A 276, 1-7.

Scherrer R. J. (2004) Phys. Rev. Lett. 93, 011301.

Schiff L. I. (1968) *Quantum Mechanics*. New York: McGraw-Hill.

Schrödinger E. (1935) Naturwissenschaften 23, 807, reprinted in English in Wheeler and Zurek (1983).

Seager W. (1999) *Theories of Consciousness.* London: Routledge.

Seager W. (2001) Panpsychism. Stanford Encyclopedia of Philosophy.

Shimony A. (1993) *Search for a Naturalistic World View, Volume II.* Cambridge: Cambridge University Press.

Smith S. (2003), Studies in the History and Philosophy of Modern Physics 34B, 261-280.

Smolin, L. (2001). *Three Roads to Quantum Gravity*. London and New York: Weidenfeld and Nicolson and Basic Books.

Spergel D. N. et al (2003) Astrophys. J. Suppl. 148, 175.

Squires E. J. (1992) Phys. Lett. A 163, 356-358.

Stapp H. (1993) *Mind, Matter, and Quantum Mechanics*. New York: Springer-Verlag.

Suarez A. (2000a) Phys. Lett. A, 269, 293-302.

Suarez A. (2000b) preprint quant-ph/0006053.

Susskind L. (1995) J. Math. Phys. 36, 6377.

Targ R. and Puthoff H. (1974) Nature 252, 602-607.

Tegmark M. (2000) Phys. Rev. E 61, 4194-4206.

Teller P. (1995) *An Interpretative Introduction to Quantum Field Theory.* Princeton: Princeton University Press.

Thomas S. (2002) Phys. Rev. Lett 89, 081301.

Tooley M. (1988) Philosophical Topics 16, 225-254.

Vallentyne P. (1997) Philosophical Studies 88, 209-219.

von Neumann J. (1955) *Mathematical Foundations of Quantum Mechanics.* Princeton: Princeton University Press.

Wackermann J., Seiter C., Keibel H. and Walach H. (2003) Neuroscience Letters 336, 60-64.

Wald R. M. (1994) *Quantum Field Theory in Curved Spacetime and Black Hole Thermodynamics.* Chicago: Chicago University Press.

Weinberg S. (1992) *Dreams of a Final Theory.* New York: Pantheon Books.

Weinstein S. (2001) in *Physics meets Philosophy at the Planck Scale.* Callender C. and Huggett N. eds. Cambridge: Cambridge University Press.

Wheeler J. A. and Zurek W. H. (1983) *Quantum Theory of Measurement.* Princeton: Princeton University Press.

Wigner E. P. (1967) *Symmetries and Reflections.* Bloomington and London: Indiana University Press.

Wootters W. K. and Zurek W. H. (1982) Nature 299, 802.

Zurek W. H. (1998) Phil. Trans. Roy. Soc. Lond. A 356, 1793-1820.

Lightning Source UK Ltd.
Milton Keynes UK
UKHW031810211022
410881UK00009B/559